P9-DNN-723

7. Friherre **Nils Reuterholm**, Landshöfding öfver Nerike och Wermland
Commendeur af Nordstjerne Orden.
Åminnelse Tal öfver Honom är hållit af Herr
von Dalin.
1756
d. 3 Dec.

8. Grefve **Carl Johan Cronstedt**, til slut President i k. Cammar Collegio,
Commendeur af Nordstjerne Orden, m. m.
1777
d. 9 Nov.

9. Grefve Herr **Augustin Ehrenswerd**, til slut Fält Marschalk, samt Ridd. och
Commendeur af Kongl. Majtz Orden.
Åminnelse Tal öfver Honom är hållit af Herr v. Arbin.
Secreterare i Vet. Acad. fr. 1740½ til 1740 ¾.
1772
d. 4 Oct.

10. Herr **Anders Johan Nordenberg**, adlad Nordenschöld Landshöfding + 1763
Ridd. af Svärds Orden. Har ingifvit 4 Rön.
öfver Honom är Åminnelse Tal hållit af Herr Kryger.
d. Julii

11. Herr **Christopher Polhem**, Commerce Råd, Commend. af k. N. O. 1751
Vald vid Acad. första Sammankomst d. 2 Junii 1739.
Ingifvit til Academiens Handlingar 20 Rön.
Var Præses för månad Jul. Aug. Sept. 1744.
Åminnelse Tal öfver Honom är hållit af H. Klingenstierna.
Des utom har Academien låtit öfver Honom prägla en
Skåde-penning.
13 augusti

12. Herr **Anders Celsius**, Astronomiæ Professor i Upsala. 1744
Vald vid k. Academiens första Sammankomst, 22 Jun. 1739.
Af des Rön finnas 20 införde i Handlingarna.
Åminnelse Tal öfver honom är hållit af Frih. von Höpken.
den 25 Ap.

13. Baron Herr **Daniel Tilas**, Bergs Råd och Landshöfding samt Commendeur af
Nordstjerne Orden.
Har ingifvit IV Rön, hållit två åminnelse Tal och
två gånger varit Præses. Åminnelse Tal öfver honom
är hållit af Herr Sandds.
1772
d. 27 Oct.

14. Herr **Olof Sandberg**, Regerings Råd. 1750
Vald d. 16 Junii 1739.
Af Honom finnes ett rön uti Handl.
uti apr.

SCIENCE IN SWEDEN

SCIENCE IN SWEDEN

THE ROYAL SWEDISH ACADEMY OF SCIENCES 1739–1989

TORE FRÄNGSMYR

Editor

SCIENCE HISTORY PUBLICATIONS, U.S.A.

1989

First published in the United States by
Science History Publications, U.S.A.
a division of
Watson Publishing International
Post Office Box 493, Canton, MA 02021
© The Royal Swedish Academy of Sciences 1989
Bidrag till Kungl. Svenska Vetenskapsakademiens
historia XXII

Library of Congress Cataloging-in-Publication Data
Science in Sweden: the Royal Swedish Academy of
Sciences, 1739–1989
    Tore Frängsmyr, editor.
        p. cm.

    Includes index.
    ISBN 0-88135-092-3
    1. Kungl. Svenska vetenskapsakademien—History.
2. Science—Sweden—History. I. Frangsmyr, Tore,
1938–
Q64.S937S25 1989
509.485—dc19   89-5931
                                            CIP

Contents

Abbreviations

K. or Kungl.	Kungliga (Royal)
KVA	Kungl. Vetenskapsakademien (the Royal Swedish Academy of Sciences)
KVA Archives	The Archives of the Academy, now located in the Center for History of Science in the Academy's main building
KVA Bidrag	Bidrag till Kungl. Svenska Vetenskapsakademiens historia, Vol, 1
KVA Historia	Sten Lindroth, *Kungl. Vetenskapsakademiens Historia, 1739–1818,* 2 parts in 3 vols. (Stockholm, 1967).
KVA Protokoll	Minutes of the meetings in the Academy, kept in the Archives of the Academy
KVAH	*Kungl. Vetenskapsakademiens Handlingar* (the Proceedings of the Academy)
KVAÅ	*Kungl. Vetenskapsakademiens Årsbok* (the Yearbook of the Academy)
KVAÖ	*Kungl. Vetenskapsakademiens Öfversigt* (Survey of the Academy's activities)
UUB	Uppsala Universitetsbibliotek (Uppsala University Library

Preface

WHEN an academy reaches its 250th anniversary, thoughts naturally turn to writing its history. But such a long span of time, filled with varied activities, is difficult to describe in a limited space. Earlier histories in Swedish have dealt with only part of the Academy's long history. The best of them, Sten Lindroth's magnificent three-volume work (1967), covers only the period 1739–1818. The sort of polyhistorian who alone could do justice to the Academy's subsequent history has proved impossible to find.

The result is therefore this collection of essays, in which we have chosen with the help of various writers to analyse some of the most important activities in the Academy's history. This means that it is not a fully comprehensive history, and that much could obviously be added; nevertheless, we hope that this book will indicate the significance of the Academy to the development of science both within Sweden and in Sweden's relations with other countries. Many of the essays overlap, but this may be an advantage, if, as can happen, the essays are read separately and not as parts of a single account.

The editorial committee responsible for planning the book has consisted of professors Gunnar Eriksson, Tore Frängsmyr, Tord Ganelius, and Inge Jonsson. The articles have been translated with the aid of a number of persons. Elisabeth Crawford wrote her article in English; the essays of Sverker Sörlin and Bosse Sundin have been translated by Stephen Fruitman, and the other articles by Bernard Vowles. In the final stages of the work, I have received much help in the preparation of illustrations by Dr. Sverker Sörlin and librarian Christer Wijkström. Arne Haglöf has prepared the index of names. I wish to thank all those who have been involved for their efforts.

Except as noted, illustrations are from the collections and archives of the Royal Swedish Academy of Sciences, Stockholm.

T.F.

The Academy's first seal, made by Daniel Fehrman in 1742. It shows the Polar star and the national coat of arms (the Three Crowns) above a collection of scientific instruments.

TORE FRÄNGSMYR

Introduction:
250 Years of Science

SEEN from an international perspective, the founding of the Academy of Sciences in Sweden in 1739 was a typical event of its time, but it also had its specifically Swedish causes. The nations with the most vigorous intellectual life had led the way. The first academies of science had appeared in Italy during the first half of the seventeenth century, to be followed by English, French, and German counterparts. The Royal Society of London (1660) wished to realize Francis Bacon's vision of a scientific community devoted to gathering useful knowledge. In Paris l'Académie Royale des Sciences was formed in 1666, under somewhat different circumstances, but with similar aims. Die Königliche Preussische Akademie der Wissenschaften was founded in Berlin in 1700 on the initiative of Leibniz, who hoped for great things from science in the future. Leibniz was also responsible for the drawings for the premises of the Imperial Academy of Sciences in St. Petersburg (1725), which was planned by Peter the Great and built by Catherine I.

All this was known in Sweden, which was certainly a great European power at the start of the eighteenth century, but which gradually slid into political chaos and accelerating economic decline as a result of the wars of Charles XII. Modern science had arrived in Sweden in the second half of the seventeenth century, principally in the form of Cartesian physics, and had managed painfully to free itself from the bonds of the Church. During the first decades of the eighteenth century, new scientific ideas were aired in a more relaxed atmosphere.

The first scientific society, known as the Collegium curiosorum, was formed in Uppsala in 1710–1711. Among its most active members, were Eric Benzelius the Younger, polymath and linguist, later an archbishop, and his brother-in-law Emanuel Swedenborg, natural scientist and engineer, later famous as a prophet and theologian. At the center of the company, was the technical genius Christopher Polhem, whose inventions the members wished to make known to a wider public. Swedenborg visited England, from where he was able to report on interesting developments in science; after his return he became the editor of a journal named *Daedalus hyperboreus* (1716–1718). Gradually the enthusiasm of the little circle cooled, but in 1719 the society was reconstituted under the name of Bokwetts Gillet; in 1728 it was granted the royal consent and renamed Societas regia literaria et scientiarum Sueciae. The record of its proceedings was given the name *Acta literaria et scientiarum Sueciae.* For a while the society's secretary was Anders Celsius, and it was through the society's efforts that the young Carl Linnaeus (later von Linné) made his first celebrated journey to Lapland in 1732.

Despite its name, this first academy never came to be seen as a national concern. It remained something of an Uppsala University coterie. The need for an academy of science to represent the whole nation therefore grew more pressing. The year 1739 was suitable for founding such an academy from several points of view. A new political party had just assumed power in the Riksdag (Parliament). This party—known as "the Hats" to distinguish it from its opponents, who were referred to as "the Caps"—pursued mercantilist policies that included the increasing of exports and the reducing of imports. The country was still in a state of economic ruin, and every effort would now be made to remedy the situation. Modern science could probably be an important factor in this process, whether in finding new useful plants or in starting industrial manufacturing.

The period between the death of Charles XII in 1718 and Gustaf III's coup d'état in 1772 is often called the Age of Freedom. By this is meant freedom from monarchic absolutism, not the freedom of the people in any modern sense. But a kind of general feeling of freedom was nevertheless characteristic of this epoch. New winds were blowing across the country. Political power now lay with the estates of the Riksdag. "The Hats" encouraged economic thinking in every field, from agriculture to manufacturing. The utilitarian was given priority whatever the cost, optimism and faith in the future were the philosophy of the day. The trading of the Swedish East India Company brought money and exotic goods into the country. Along with the expansion in foreign trade, the intellectual exchange with the outside world also became livelier; scientists, scholars, and writers went abroad and came home with new impressions and the stories of their travels. At home major economic projects were started—money was pumped into grandiose schemes for such activities as textile manufacturing and silkworm farming. Science was harnessed to economic utility, and in 1750 a committee proposed terminating the professorships in oriental languages and Latin verse at Uppsala University in favour of new chairs in physics and chemistry.

The contrast between the old and the new was not always expressed as starkly as this, but the example shows the mood of both politicians and intellectuals. What was of value to the country counted for more than the individual's need for education and culture. The old professorships in oriental languages and Latin verse were not withdrawn, but the proposed new posts in physics and chemistry were duly introduced. There was no point in waiting; there was no time to be lost.

The Founding of the Academy

This was the spirit in which the Academy of Sciences was born. Universities functioned as teaching establishments; the professors were not expected to do any research. Academies had shown new ways of acquiring knowledge that was practical and useful. The Royal Society's insistence on its practical purpose is well known; it refused to concern itself with anything that it regarded as metaphysics.

The practical and economic view of the purpose of science therefore united the six scientists and politicians who met at the Palace of the Nobility in Stockholm on June 2, 1739 to form the new Academy. The group was made up of the botanist Carl Linnaeus, the engineer Mårten Triewald, the factory owner Jonas Alström (later Alströmer), the Hat politicians Anders Johan von Höpken and Carl Wilhelm Cederhjelm (absent on this occasion), and Sten Carl Bielke, who was both a politician and a good amateur botanist. The aim of the new Academy would be to promote and spread all possible useful branches of science in Sweden. This emphasis was so pronounced that they thought of calling the new society an "economic academy." This was not done, but that it was even considered clearly indicates the direction in which their ambitions lay.

The first meeting took an important decision. Every quarter the Academy would publish its proceedings as a journal, *Vetenskapsakademiens Handlingar,* which would be written in Swedish, not Latin. This was meant to reinforce the intention of bringing science to the people. Anyone who could read would be able to share the latest knowledge, whether in agriculture, handicrafts, or industry. The Proceedings contained "observations" *(rön),* i.e., practical experience and advice, and also, more occasionally, theoretical essays. The subjects were often concrete ones: how to exterminate pests, how to tar roofs, how to improve the harvest with new tools and methods, or how best to draw beer and wine from the barrels in the cellar.

A symbol of the Academy's endeavours was provided by the frontispiece of the first volume of the Proceedings: an illustration of an old man digging a hole to plant a tree. Above his head may be seen a banner proclaiming "For Posterity." With time and in the hands of other engravers, the face of the old man changes: sometimes he looks like a vigorous middle-aged man, sometimes like a veteran all-in wrestler, but he is always present. He plants for coming generations, which is precisely what the patrons of science in the new Academy themselves wished to do.

The Academy was evidently a success right from the start. One reason for this was, of course, the political support it enjoyed, even if this did not bring any actual financial advantage. Höpken, who acted as secretary for the first few years, had many contacts. He arranged a royal charter and framed statutes that also obtained the king's approval. Being able to call itself "royal" gave the Academy status, if not money. The Swedes' immediate organizational model was the Royal Society of London; they wanted the Academy to have independent status and to decide for itself the form its activities would take. The idea of copying the French Académie des Sciences in having permanently employed members certainly cropped up from time to time, for example when Pehr Wilhelm Wargentin was secretary, but this was never put into practice.

The activities of the Academy were initially of three different kinds: meetings, lectures, and the publication of the Proceedings.

Carl Linnaeus (1707–1778), one of the founders of the Academy and its first President.

Mårten Triewald (1691–1747), inspired by his visit to London and the example of the Royal Society.

To these was soon added the Academy's own program of research, primarily in astronomy. Aspirations were high. Meetings were to be held at least once a week; there were often two meetings a week during the first years, although they became somewhat less frequent later. A president was appointed for each new quarterly period, and he had to give an address when he retired. These presidential addresses might review the state of a particular area of science or propose a program of action in that sphere, or they might present a program of a more general nature. They therefore tell us a lot about the scientific questions that were topical at any given time. Other important series of addresses yield further information because there were both inaugural speeches by newly elected members and commemorative speeches to honor members who had died. The latter, which were inspired by the French *éloges,* were often based on the deceased member's autobiography; the secretary had no qualms about reminding older members to submit their curriculum vitae in good time.

The question of allocating the members to specific sections was seen as urgent from the start. The Academy looked at how foreign predecessors worked and saw that the there was a strict classification of members, at least in Paris, Berlin, and St. Petersburg. The Royal Society also had a system of classification although it was never properly applied in practice. By 1740 the Academy had devised its first classification system, the work of the mineralogist Daniel Tilas. There were five sections:

    1. Astronomy.
    2. "Elementa," i.e., experimental physics, mathematics, mechanics, architecture.
    3. Natural history, including fishing, hunting, and silkworm farming.
    4. "Artificialia," i.e., economics, industry, trade, and medicine.
    5. "Lingua," or the Swedish language.

In practice, however, there was still no real classification. For one thing, the boundaries between the sections were obviously indistinct; for another the sections were seen more as interest groups, which led some members to assign themselves to more than one section. Incoming observations were to be referred when received to the appropriate section, but no definite organization was ever introduced. "Section" merely became a convenient way of summarizing the members interested in a particular field.

The statutes said nothing about how many members the Academy was to elect. By 1750 there were about 100, and it began to be felt that this might be a suitable number. Without being inflexible, the Academy followed this guideline for the rest of the eighteenth century. Eighty percent of the members came from one of two distinct groups—aristocrats and bureaucrats and university professors; but pharmacists, priests, doctors, and officers were also represented. That the nobility played an important part has already been indicated. Persons of power and influence were often elected solely for the benefit that this might bring to the Academy, and no one saw anything improper in this. Linnaeus wrote to Wargentin quite openly on the subject. He said that the Academy consisted of two kinds of members—working members and patrons: the former reported on their scientific findings; the latter encouraged and promoted and might be very valuable to the Academy, since they were the "mainsprings of the whole kingdom." Höpken, who was himself one of these important gentlemen, also said that birth and rank were weighty qualifications for membership; all academies had their honorary members: "They shine, they may be useful but they never do harm."

One lady was also elected to the Academy, Countess Eva Ekeblad (née De la Gardie). When she was proposed, the precedent of Madame Châtelet, Voltaire's mistress, who had been elected to the Academy of Bologna, was quoted. Countess Ekeblad was elected in 1748, after submitting a short article on the production of starch and powder from the potato; it was well received and was included in the Proceedings. Later she sent in an article on how soap could be used to bleach cotton. Eva Ekeblad was the wife of Councillor of the Realm Claes Ekeblad, who was also among the aristocrats in the Academy, and this no doubt smoothed the way to her election. But it should also be noted that the countess never attended a meeting, so perhaps her membership was regarded as purely honorary.

There were other people, too, on the periphery of the Academy. It was soon decided to set up a section for "subject novices," i.e., a kind of apprentice class in which wisdom and scientific understanding could be acquired within the framework of the Academy. But geographical considerations were themselves enough to prevent such training from actually materializing, and the designation came rather to denote a kind of associate or corresponding member. As an incentive even these modest nominations had a very significant effect. Priests, doctors, and skilled craftsmen in the provinces were invariably delighted at this notice from the exalted heights of the Academy in Stockholm. Not until near the end of the century was the practice of appointing subject novices discontinued.

These activities show, however, that the Academy really did aspire to educate the people. Contact with the people was important, through both the printed word and public lectures. When Johan Carl Wilcke was engaged as assistant physicist in 1759, his most important task was to give public lectures.

Foreign Relations

But the Academy did not regard its work as only for domestic consumption—it was also anxious to be part of the learned world of Europe. In his long period as secretary, Wargentin skillfully managed to develop a large network of contacts. He became something of a clearing house for the far-flung stations of science, both in Sweden and even in other countries. Naturally correspondence as such was the most important means of communication in an age when there were limits to how much travel was feasible. The predecessors of the Royal Society had realized this, more than a century earlier. When Samuel Hartlib set up his Office of Address in London in 1646, his idea was to establish a kind of international correspondence college for scientists. This was precisely the intention that Wargentin was to put into effect.

Sten Lindroth, who has written the most comprehensive history of the Academy up to the accession of Berzelius in 1818, describes Wargentin's diligence and efficiency. During his time as secretary, he wrote continuously. A register of his correspondence covering the years 1759–1783 shows that during this period he sent 1,193 letters abroad, a rate of about one a week. He himself also maintained that regular correspondence had succeeded in creating "some communication or a kind of brotherhood between our academy and some of the most famed overseas." A scientific academy without correspondence had little life, he believed.

Collaboration with the Royal Society in London was put on a more formal basis in 1750, when Wargentin and his English colleague Cromwell Mortimer agreed to exchange publications. In the following year, co-operation with l'Académie Royale des Sciences began, the link here being Wargentin's close friendship with his fellow astronomer Joseph Nicolas de l'Isle. Many of these contacts had been nourished by Swedish travellers and diplomats. Personal communication was the vital factor. London was always being visited by young students and scientists, such as the astronomer Bengt Ferrner, who later wrote an unrivalled account of his lengthy tour. Also in London was the well-known clockmaker John Ellicott, who acted as Swedish commissionaire, and he was, of course, an additional asset to all visiting Swedes. Anders Celsius saw a lot of Ellicott while he was in London in 1735–1736, and when it came to buying scientific instruments, Ellicott's assistance was invaluable. In 1754 Ellicott became the second Englishman to be elected to the Academy of Sciences.

In France the first links between the academies of the two countries were forged with the aid of the Swedish minister in Paris, Carl Fredrik Scheffer, who was influenced by the physiocratic ideas that he later brought home to King Gustaf III. But the role played by Ellicott in London was played in Paris by the chaplain to the Swedish legation, Frédéric-Charles Baer, actually a native of Alsace, but so long and loyally in Swedish service that he was eventually ennobled. Baer was at home in scientific and philosophical circles; he knew Voltaire and Diderot and was himself a corresponding member of l'Académie Royale des Sciences. He translated Swedish publications into French, looked after visiting Swedes, and was an active intermediary between the two academies.

On the other hand, ties with Berlin's academy of sciences, surprisingly enough, never became very close. Although Linnaeus was elected as a foreign member, as later was Daniel Melander-hielm, other contact was sporadic. Nor was there any exchange of publications. However, in the 1770s Wargentin began corresponding with his colleague Jean Bernoulli, who performed a number of services for the Swedes. The Academy had much closer relations with Die Sozietät der Wissenschaften in Göttingen. This academy was founded at the new university there at the instigation of Albrecht von Haller. Haller corresponded for many years with the physician Nils Rosén von Rosenstein and, until they fell out, with Linnaeus. Haller remained in continuous touch with Wargentin by letter, was elected as a matter of course to the

Count Anders Johan von Höpken (1712–1798), the first Secretary of the Academy.

Jonas Alströmer (1685–1761), factory owner and particularly interested in
the new sciences.

Academy, and had articles published in the Swedes' Proceedings. When Haller returned to his native Switzerland, the flow of letters in both directions continued unabated. Haller also learned Swedish, which enabled him to follow events in Sweden. His regular reviews of Swedish work, including published addresses to the Academy, appeared in *Göttingische Anzeigen.* Many others in Göttingen helped the Swedes to obtain information from the world of learning or to buy newly published books, among them being the expatriate Swede Johan Andreas Murray and the erudite August Ludwig Schlözer.

There was also some co-operation between the Swedes and the academy founded in St. Petersburg. There were even offers of professorships at St. Petersburg to Pehr Kalm, Pehr Adrian Gadd, and Daniel Solander. The young Finn Erik Laxman took part in an expedition to Siberia, Mongolia, and China, returning with such an impressive collection of material that he was appointed professor of economics and chemistry by the St. Petersburg academy in 1770. Another Finn, Anders Johan Lexell, was recruited from Åbo (Turku) as an assistant. From Sweden itself it was, as usual, Wargentin whose regular correspondence kept the two academies in touch, especially after he had found a like-minded spirit in St. Petersburg in Johann Albrecht Euler, himself a mathematician and the son of the famous Leonhard Euler.

One well-established method of cementing foreign relations was, of course, to elect foreign colleagues to the Academy as members. At first there seem to have been no firm rules for this practice. The Academy's first move in this direction was to elect 25 foreign members at one go in the 1740s, including such names as Maupertuis, de l'Isle, Euler, Bradley, and Daniel Bernoulli. Strange to relate, only 15 appear to have acknowledged the honor; nothing was heard from the other ten, so in a sense their election was never formally ratified. More foreign members were elected later, such as de Jussieu, Réaumur, Musschenbroek, Sauvages, and Gmelin.

The original idea was that each foreign member would be "looked after" by a Swedish colleague, and that the pair would maintain regular contact by letter. However, this practice never became established. Some foreign members hardly ever wrote; others, such as Paolo Frisi in Milan and Charles Bonnet in Geneva, made a valuable contribution to the work of the Academy.

Distribution of the Academy's publications abroad naturally

depended entirely on the existence of local contacts. Sometimes these were foreign members, sometimes students who knew enough Swedish to be able to translate or who obtained translations from others. In those days copyright and similar legal niceties were not taken too seriously. Texts were translated, abridged, or quoted with little scruple. For example, it was not uncommon for Swedish articles to be retranslated from German, increasing the risk of inaccuracies. There were other hazards in this casual attitude to the written word. Wargentin found to his surprise that letters he had written to Albrecht von Haller had been reproduced in full in the latter's *Epistolae* (1774), a compliment that he tactfully hoped would not be repeated. Enthusiasm for spreading news from the field of science could sometimes overstep the bounds of propriety.

Decline and Revival

When Wargentin died in 1783, it had a profound effect not only on the Academy, but also on Swedish science in general. A period of decline set in, which cannot be blamed on a single individual or a single event and is therefore difficult to explain.

By the mid-1780s, the great names of Swedish science had gone: Linnaeus, Bergman, Wargentin, Wallerius, and, in 1786, Scheele. Celsius and Klingenstierna had died long before. The elder generation had departed, and of the new one, there was no sign. The decline had actually set in much earlier. Even in the 1760s, the glamor of the new science was beginning to fade: the economic expectations had not really been fulfilled. When the novelty wore off, other amusements began to come into fashion: the king and the court turned to ballet, opera, and belles lettres. Serious science was considered boring. Occultism and currents of mysticism began to attract attention, and not far from the royal palace at Drottningholm, fortune-seekers began trying to produce gold. Then came Romanticism, questioning the whole spirit in which the Academy had worked during the eighteenth century.

All these are factors that need to be mentioned when this marked change in the cultural climate is being considered. Nevertheless, they do not go very far by way of explanation. It is as if the trend of the times veered off on a new course without anyone really knowing why, as if people had had their fill of a certain kind of culture and sought something new. The Academy of Sciences was a national body; it was supported by all. But it also stood for a certain kind of culture: our "heathen academy," Bishop Andreas Rhyzelius called it. The Academy had a rational program; it glorified science; it devoutly preached economy and material benefit. The new tones that were being heard elsewhere in Europe said that it was time for imagination and joy, song and music.

Some of the earlier generation survived a little longer, of course, even if it became harder for them to be heard. After Wargentin, Johan Carl Wilcke was elected secretary in 1784. He was a leading scientist and in every way a likeable man, but he lacked Wargentin's organizational skills. He was not capable of becoming the unifying force that the Academy needed at this moment more than ever. When Wilcke died suddenly in 1796, aged 63, the 70-year-old astronomer Daniel Melanderhielm was elected to succeed him. The inventive Melanderhielm had never really made a name for himself, and he did not achieve much for the Academy in general before he retired in 1803. Nevertheless, he deserves recognition as the driving force behind the expedition to measure a degree of the meridian in 1802–1803, and as a result of his initiative, the credit for the project became the Academy's. The leader in this field was Melanderhielm's younger colleague Jöns Svanberg, who was also one of the vice-secretaries appointed to give the ageing Melanderhielm some assistance.

A small flock needs a strong shepherd, and a frail academy certainly needs a strong secretary. The decline in the late eighteenth and early nineteenth centuries expressed itself in many ways. The quality of the articles in the Proceedings deteriorated, meetings were held less frequently, and attendance decreased; sometimes, only 15 or so members might be present for votes on important issues. The gap left by Wargentin took a long time to fill. Melanderhielm was, as we have seen, already an old man when he became secretary. He in turn was succeeded by the young physicist Carl Gustaf Sjöstén, whose nervous temperament and lack of business acumen led to failure both in his work for the Academy and in his private life. He resigned in 1808, and Jöns Svanberg was elected unanimously as his successor. But he brought little improvement. Svanberg was in many respects an excellent man, but after little more than two years, he too resigned to take a chair in mathematics at Uppsala University. His reason for leaving his post is not known, but Lindroth assumes that he thought he would have more time for research with a professorship. After Svanberg the candidates for the secretary's position were the botanist Olof Swartz and the chemist Jacob Berzelius. The vote went in favor of the former by 21–17: Swartz was older and had a sound international reputation as a scientist. In his youth he had distinguished himself with his journey to the Americas in 1783–1786, returning with a large collection of botanical specimens from the West Indies. Swartz, too, would clearly have been a splendid man for the job in many ways, but he insisted on trying to combine his professorship in botany with his duties as secretary of the Academy; moreover he was the curator of the Academy's museum and was later appointed to a second professorship, at the Karolinska Institute. It need hardly be said that in the long run this unreasonable workload did no good at all to either the Academy or Swartz himself, and in September 1818 he died.

When Jacob Berzelius became secretary in 1818, the Academy entered its second great period. Admittedly, when he took up his post, it was still the age of Romantic natural philosophy, when the exact sciences were scorned, but Berzelius did not worry unduly about that. He spoke his mind about the Romantics, and his position was such that charges of materialism could not hurt him. He was energetic and a good organizer; he may also have been something of a despot, but his commitment was to the cause of science.

It was typical of Berzelius that he at once set to work on a thorough reorganization of the Academy, both its statutes and its program of work. One of his first acts was to introduce a new classification system. It was clear to him that this was one of the preconditions of efficient and purposeful activity. We have already noted that the first attempts to divide the Academy into sections were not altogether successful. Nor had the results been any better when a new attempt had been made in 1798. The explanation advanced then was that the sciences were not represented in the correct proportions. In particular there were complaints that despite the importance of agriculture there were far too few rural economists. Others were critical of the very principle of a strict division into sections, which they saw as an instrument of autocratic control. One of the critics, the royal physician and mine councillor Nils Dalberg, argued furiously against a reform. But he added that should the new classification nevertheless be introduced, he wished to be included among the agriculturalists, "in anticipation of my own early return to dust."

The Palace of the Nobility (the large building on the right) became the first home of the Academy. Engraving ca. 1780.

The new system consisted of seven sections and reflects the contemporary view of the sciences:

1. General economics and agriculture.
2. Trade and civil industry including mining.
3. External knowledge of nature and natural history, i.e., geology, crystallography, botany, and zoology.
4. Internal knowledge of nature and experimental philosophy, including chemistry, physics, and meteorology.
5. Mathematical sciences, primarily geometry, astronomy, and "physica mathematica."
6. Medical practice, including veterinary medicine and anatomy.
7. The literary section, "literature, the history of the world and of learning, languages and other useful and graceful sciences."

On paper this reform indicates a move towards economics, agriculture and trade, and a contraction of the theoretical sciences (astronomy, for instance, had previously had a section of its own, whereas it was now placed under mathematical subjects). In practice, however, the divisions still had no real significance.

In 1820 Berzelius proposed a new classification. The trend is clear. Titles such as trade, agriculture, and industry disappear or are subsumed under the heading of "economic sciences." Berzelius wished to give the Academy stability by introducing theoretical designations and then following the disciplinary divisions of the universities. In content the implication of this was that science, rather than practical commerce, was what was wanted. There were now nine sections:

1. Pure mathematics.
2. Applied mathematics.
3. Practical mechanics.
4. Physics.
5. Chemistry and mineralogy.
6. Zoology and botany.
7. Medicine.
8. Economic sciences.
9. General scholarship.

A comparison between the two classifications shows that Berzelius anticipated a new age; his reform would survive into the twentieth century. The subject of the division of the Academy into sections may appear a tedious and formal aspect of its activities, but it undoubtedly reflects the view of sciences current at a particular time.

Berzelius did much to make exact science respectable and culturally acceptable. For example, he—albeit rather reluctantly—allowed himself to be elected as one of the 18 members of the Swedish Academy (of literature and language). The two academies drew steadily closer and had other members in common, namely the botanists Carl Adolph Agardh and Elias Fries. Alongside Berzelius' own speciality, chemistry, other disciplines also flourished, both within and outside the Academy of Sciences. Botany probably had the easiest ride. Thanks to Linnaeus, it had become something of a national science, and admiration of Linnaeus became almost a cult for the Romantics.

Berzelius, Agardh, and Fries all played important parts in the popularization of science. Berzelius had written chemistry textbooks before he became a chemist of international eminence, and he sat on the Government Commission on Education and accepted other civic duties. Agardh was a specialist in algae and also a remarkable economist. He, too, was a member of the Commission on Education; he expounded the teachings of Montesquieu; and he argued for the greater freedom of women in society. He was elected to the Academy of Sciences as early as 1817. Gradually his theological interests gained the upper hand, and in 1835 he was made a bishop. Agardh's career shows that a scientist could still play many different parts in the life of the country, despite increasing specialization.

His colleague Elias Fries was another example of the cultured scientist. He became a member of the Academy in 1821 and is regarded as possibly second only to Linnaeus as a botanist, with fungi his speciality. In 1831 he was distinguished by the Academy, which later commissioned him to make an important contribution to popular education, namely, a large book of plates on the fungi of Sweden, both edible and poisonous, *"Sveriges ätliga och giftiga svampar"* (1860–1866). This was probably more influential than many written texts. But, of course, it must not obscure the fact that Fries was also highly regarded as a writer, particularly for his *Botaniska utflykter* (3 vols., 1843–1864). It should also be mentioned that, like Berzelius, Fries was apt to speak of science as an aspect of culture. Another scientist of influence was the versatile

Sven Nilsson, who was a zoologist, geologist, and archaeologist, became a member of the Academy in 1821, and was the keeper of its collections from 1828–1831. He studied fossils and the glaciation of Scandinavia and published works on its indigenous peoples.

The celebrated physicist Anders Ångström also did useful work for the Academy, even if the scene of most of his activity was Uppsala University. He was assistant at the Academy's observatory for a year (1842–1843) before he moved to Uppsala. The Academy entrusted him with the task of editing the data on terrestrial magnetism and meteorology obtained during the circumnavigation of the globe by the frigate *Eugénie* in 1851–1853. Ångström published many of his most important experiments in the Proceedings of the Academy.

Berzelius was, of course, the dominating figure in the Academy, but his contribution both raised the status of the Academy directly and stimulated his colleagues in other spheres.

A Faculty of Science

Great expeditions were a prominent feature of the second half of the nineteenth century, but the origins of this activity actually go back further. Berzelius ensured that travel grants were introduced to explore Sweden itself. Young students and scientists were able to discover different parts of their own country, following in the footsteps of Linnaeus. The voyage of the *Eugénie* extended the horizons.

The major polar expeditions began, in fact, as geological field studies and ended in projects that attracted international attention. A unifying force throughout this period may be seen in the person of Sven Lovén. He took part in the first polar expedition in 1837 and lived until 1895. It was through Lovén that his pupil Otto Torell became interested in Arctic molluscs and began his study of glaciation. Torell fitted out the first large expeditions and passed on the benefit of his experience to his assistant Adolf Erik Nordenskiöld, who brought international renown to Sweden and the Academy of Sciences. When Lovén died, he had been a member of the Academy for 55 years.

To the public at large, however, the Academy came first and foremost to be identified with Nordenskiöld. As curator and professor of the Museum of Natural History he was an employee of the Academy, and only three years after taking up employment he was elected a member (1861). The Museum was housed in the Academy building, the Westman House on Drottninggatan that had been purchased in 1828 by Berzelius, and Nordenskiöld also had his living quarters there. In *Nya Stockholm* (1890), Claës Lundin depicts the scene in Nordenskiöld's study, where the great man was perusing old maps:

"It is in this modest low-roofed dwelling that Baron Nordenskiöld, one of the most widely known learned men of our time, has his quiet and agreeable home, where, with the exception of the short time in the summer that he spends on his property at Dalbyö in Södermanland, he rests with his wife and children between long and perilous excursions and considers to which quarter next to steer his course, all the while completing some new scientific work and discharging his duties as the curator of the mineralogical collections at the Museum of Natural History."

All the expeditions, both Nordenskiöld's and others, had as their purpose observation, description, and gathering natural history specimens of all kinds: plants, animals, fossils, stones, etc. During the nineteenth century, the old collections were structured and the Museum of Natural History was established and quickly expanded. For a long time, the Museum was an institute of the Academy. A similar position was occupied by the Central Meteorological Office, which was an autonomous establishment but operated under the Academy's supervision. Erik Edlund, the Academy physicist since 1850, had proposed in 1856 that the Academy should organize a program of meteorological activity: all over the country, stations or reporters, would take observations that would be sent to the Academy. Edlund himself led this work until 1873, when the meteorological office was set up, and Robert Rubenson became its director. At that time there were approximately 40 weather stations in the country, but by 1890 the number was already 450.

Some of these observations were carried out at lighthouses and supervised on behalf of the Academy by Axel Erdmann, a geologist who became a member in 1846. Erdmann later became the head of the Swedish Geological Survey, which was set up in 1858. He was succeeded in this capacity by Otto Torell (1871), who already had strong ties with the Academy. Together with the Academy, he had, as has been mentioned, planned and carried out the first major polar expedition; he had been given a large travel scholarship by the Academy; and it was in the Proceedings of the Academy that he had presented his epoch-making observations and conclusions on the effect of the ice age. For the rest of his life, Torell remained in close personal contact with Nordenskiöld.

Another institution with which the Academy worked closely was the Karolinska Institute, Stockholm's center for medical studies. Many of its professors did a great deal on the Academy's behalf: Gustaf Retzius, for example, was a histologist, anatomist, and anthropologist, who was elected in 1879 and in the years that followed received numerous prizes and distinctions from the Academy. He also wrote the biographies of several of its most prominent members, including Linnaeus, Sven Nilsson, and A.E. Nordenskiöld.

It can confidently be stated that the Academy was actively involved in every scientific activity that took place in Stockholm in the nineteenth century. Since for a long time the capital had no university, the Academy filled the role of a kind of faculty of science, just as the Karolinska Institute formed a medical faculty. And when in 1878 Stockholm finally did acquire a university college or *högskola* (comparable to a university but without the full range of faculties), members of the Academy played a leading part in helping the new seat of learning establish itself. One of these was Gösta Mittag-Leffler, a strong and controversial personality, professor of mathematics, and twice rector of the college. Mittag-Leffler donated his house, his mathematics library, and his fortune to a foundation, the management of which was entrusted to the Academy.

One of Mittag-Leffler's more noteworthy initiatives was to bring Sonja Kovalevsky to Stockholm in 1883, to be appointed as professor of mathematics the next year. Her talent was undeniable, but the status of women was such that they could not automatically study or obtain appointments on anything like an equal footing with men. The new *högskola* was not bound by old traditions, however, and Mittag-Leffler was an enterprising man. He quickly drummed up donations and took the board of the college by surprise. He had less success with the Academy. Only a year after her professorial appointment, Mittag-Leffler tried to have Sonja Kovalevsky elected: she was formally proposed by Carl Malmsten, his colleague in Uppsala, and two more mathematicians seconded the proposal. However, the opposition was too strong. The opponents claimed that the statutes referred only to "men," although others maintained that "man" now meant "person." The permanent secretary, Georg Lindhagen, spoke against the proposal and the Academy agreed with him. It was decided by 28 votes to 13 that a woman could not become a member of the Academy of Sciences.

The first woman to be elected in modern times was in fact Lise Meitner, who came to Sweden in 1938. She was elected as a foreign member in 1945, but after becoming a Swedish citizen in 1949, she was transferred in 1951 to the Swedish division. After Lise Meitner, the next women were the American botanist Katherine Esau in the foreign class (1971) and the astronomer Aina Elvius in the Swedish one (1975). As it celebrates its 250th anniversary, the Academy has eight female members.

The Science City

As the högskola in Stockholm grew, the importance of the Academy as a center of research declined. Several of its institutes were incorporated in the högskola, which was formally designated a university in 1960, and became part of its faculty of science. But the Academy continued to function as a lively forum for scientific debate and a co-ordinator of national scientific initiatives. Examples that come readily to mind are its work for environmental protection and its role in the creation of Sweden's national parks.

Another landmark in the Academy's modern history is its move in 1915 to its present premises in the Frescati area just north of Stockholm. A brief recapitulation of the various buildings that the Academy has occupied may be appropriate here. From its foundation in 1739 until 1764, the Academy met in the Palace of the Nobility in Stockholm, which was natural enough as many of its leading members belonged to the aristocracy. For a short while afterwards, the Oxenstierna House at Storkyrkobrinken in the Old City was used, as was Greve Pehr's House on Helgeandsholmen. In 1779 the Academy moved into the Lefebure House on Stora Nygatan, which had been purchased two years earlier, and it remained there until 1828 when Berzelius obtained the Westman House on Drottninggatan.

This home, too, eventually became too small, largely because of the rapid increase in the size of the Academy's natural history collections. After much discussion the decision was taken to have

The main building of the Academy, opened in 1915. The smaller building on the right now houses the Berzelius Museum. Drawing by Lars Dhejne, 1981.

new premises built in Frescati (so named after the journey of Gustaf III to Italy in 1783–1784). The result was the big main building that is used today, together with a number of smaller buildings as residences for the Academy's staff. On November 10, 1915, the Academy held its first meeting in the new building. At the same time, an imposing new edifice, with two wings, was erected for the Museum of Natural History and opened a year later. The architect responsible for both buildings was Axel Anderberg. The same area had also long been the site of the Bergius Garden, while the schools of forestry and veterinary science, the experimental grounds of the Royal Academy of Agriculture, and the Swedish Geological Survey were all close by. The presence of this group of scientific institutions led to the area being dubbed *Vetenskapsstaden* (Science City). Some of them moved to other districts in the 1970s, but their place was quickly taken by the main building of Stockholm University and most of its departments. The Academy of Sciences is therefore nearer to the university and its students than ever before.

As far as the nature of its activities is concerned, the Academy today is usually associated with the Nobel prizes for physics and chemistry (and, since 1969, the prize for economic sciences introduced in Alfred Nobel's memory). The prizes have been awarded since 1901 and have naturally affected the Academy's activities. Many members were dubious about accepting the task of awarding the Nobel Prize. They feared that this responsibility might so dominate the program that other activities would grind to a halt. With hindsight it seems clear that their anxiety was exaggerated. On the contrary, it is fair to say that the Nobel work has been a stimulating factor, which has enhanced the status of the Academy in the scientific community. It has brought the Academy into the international limelight, and has also led its members to follow the latest developments in research with even closer attention.

Another consequence has been that the function of the Academy as a prize-awarding body has inspired new donors. The Academy probably makes more awards in scientific subjects than any other body in Sweden. The most generous donation of more recent times is the Crafoord Prize, which is the largest award after the Nobel Prize. A range of other activities demonstrates that the Academy does not exist solely to award the Nobel Prizes. As various activities have been cast off to function as autonomous units, or transferred to Stockholm University, new projects have been started. At present the Academy is responsible for seven research institutes, publishes seven scientific journals, and organizes the work of about 20 national committees. Internationally the Academy is well represented in committees and research exchange programs.

A New Openness

The latest phase in the Academy's history began in the 1970s. The event that prompted it was the Academy's loss in 1972 of the exclusive right to publish the Swedish almanac, which it had enjoyed since 1747; it was thus obliged to look for other sources of income. There were still foundations and donations to rely on, and an annual grant was now approved by the Riksdag. This grant did not by any means imply that the Academy surrendered its independent status; it has continued to function as an independent body in the research community.

Nevertheless, certain changes had to be made. In many respects the Academy had become a club for elderly gentlemen. In 1973 a rule was introduced providing for the election of new members when existing ones reached the age of 65. This had a rejuvenating effect, even if members over age 65 were naturally still welcome to take part in activities.

At the same time, there was a new openness to the general public. The Academy's meetings, which are held on the second and fourth Wednesdays of each month, were divided into two parts. The closed meeting of the Academy's members in the old portrait-lined assembly room is now followed by an open session, consisting of a lecture and a discussion in the large lecture theater. Interested visitors from outside the Academy are welcome; invitations are often extended to guests who are expected to have a special interest in the subject under discussion. The evening always finishes with a Swedish smorgasbord, to which, too, guests are often invited. Over the last ten years, it has become the practice to devote the first meeting in January each year to the government budget proposals that have just been presented. The minister of research attends, often accompanied by the undersecretary and other officials from the ministry, to explain the government's proposals for research and higher education. This is followed by questions and discussion of the country's research policy. There is usually a large audience on these occasions.

The changing nature of the Academy's work over the years has naturally been reflected in changes in the classification system. We have already seen how Berzelius reorganized the categories to correspond with the research situation at the universities. Berzelius' system remained in force throughout the nineteenth century. The statutes of 1904 increased the number of sections from nine to eleven. This was done by dividing the chemistry and mineralogy section into two—one for chemistry alone and one for mineralogy, geology, and physical geography—and by giving zoology and botany their own sections. The "practical mechanics" section changed both name and position; it was renamed "technical sciences" and moved from number three to number nine. The next major change was in 1947, when "geophysics" was added as number 11. The previous final category, formerly called "general scholarship," was now retitled "for other sciences and for outstanding service to scientific research" and made number 12.

Scientific progress once more led the Academy to reorganize its sections just before its jubilee year. It was particularly felt that recent developments in biology rendered the old categories of botany and zoology inappropriate. Consideration was also given to starting a separate section for cell and molecular biology. There was also discussion of how physics, astronomy, and geology were to be classified. The fragmentation of sciences has the paradoxical effect of also increasing communication because some subjects require interdisciplinary treatment. This makes it difficult to be definite about what belongs to one discipline and what to another. The Academy eventually decided to amalgamate rather than to sub-divide, reducing the number of sections but increasing the number of members (under 65 years old) to a maximum of 161 (instead of the previous 134). The new statutes were approved in December 1988, and the sections are now as follows:

  I. Mathematics.
  II. Astronomy and space science.
  III. Physics.
  IV. Chemistry.
  V. Earth sciences.
  VI. Biological sciences.

VII. Medical sciences.
VIII. Technical sciences.
  IX. Economic and social sciences.
   X. Humanities and other sciences and outstanding
      service to science.

It will not take as long as it did in Berzelius' day for a new revision to become necessary. But the Academy has probably ensured that the present arrangement will remain viable for longer by opting for a unifying approach.

Another sign of the Academy's new openness is that it has undertaken various fact-finding tasks in connection with particular research issues. During the jubilee year, for example, a survey of the international relations of Swedish research is being carried out. This task can be invested with an almost symbolic significance. The international perspective has to be paramount in a living academy of sciences. Collaboration and projects transcending international boundaries are vital to research that aspires to be universal rather than provincial.

When the Royal Swedish Academy of Sciences was founded, Linnaeus and his colleagues wished first and foremost to stimulate the growth of science in Sweden, but they knew that the best way to do this was to establish ties with science in other countries. Although external circumstances have changed during the 250 years that have passed, this remains the overall aim of the Academy's activities.

SVEN-ERIC LIEDMAN

Utilitarianism
and the Economy

IN his presidential address to the Academy of Sciences in 1744, entitled *Tal om nyttan, som insecterne och deras skärskådande, tilskynda oss* [On the advantage we derive from the insects and from the examination of them], Charles De Geer maintained that *historia insectorum* is generally regarded as "unprofitable, of little benefit to us, and thus conditional in its very existence." But De Geer himself, whose income came from his land, was an entomologist and naturally held a different opinion. He admitted that there were sciences of greater utility than his own, but asserted that knowledge of insects can also be valuable.

The question of utility was debated continuously in Sweden during the Age of Freedom. It recurred constantly in the early history of the Academy. In one presidential address after another, different utilities were compared, and there was even a certain amount of unobtrusive rivalry between presidential addresses over their ranking order. Not everyone was as modest as De Geer, who assigned his own pet subject to the second rank.

The word "utility" undoubtedly had a wide range, as terms that become popular often do. It also included whatever was important to the preservation of good morals and godliness. The first of the good effects that De Geer attributed to the study of insects is that it gives us an idea of the vast number of strange and ingenious species that exist and thus reminds us that we have reason to "praise the most high." We find similar thoughts in many other places, reaching, perhaps, their most eloquent expression in Linnaeus' celebrated academic dissertation *Cui bono?* (1752).

But the criteria of utility are chiefly secular and first and foremost material. They are primarily, as De Geer noted with a sensitive awareness of the content of the word, those things that concern our bodies and that contribute to their maintenance, clothing, and health but also to household management in general.

The term "household management" is the direct equivalent of the Greek word *oikonomia* (economy). The question of what is useful is closely linked to the question of what is economic. Economic matters played a large, indeed at first a dominating, role in the Academy. As late as 1770, Pehr Wargentin could speak of "an economic Academy of Sciences such as ours."[1] By that time, it is true, economic discussion and economic subjects had started to figure somewhat less prominently in the Academy's presidential addresses; but the number of economic observations in the Proceedings was still large.

It will be our task to examine more closely the striking interest taken in economic problems by the Swedish Academy of Sciences. But first we must clarify the general ideo-historical background.

Economics as a Subject of Study

Economics, like politics, was considered by Aristotle to be one aspect of practical philosophy, in which form it long lived a more or less obscure life in the universities. It was as a professor of moral philosophy that Adam Smith revolutionized the science of economics at the end of the eighteenth century.

The domain of moral philosophy bordered on that of jurisprudence, and in the study of this subject, too, there was a place for economics. Here the emphasis was on economic jurisprudence, or the theory of the law on taxation, trade, and other down-to-earth subjects.

The basic Greek meaning of the word "economy" survived until the end of the eighteenth century or longer. Economy was in other words household management in a literal sense, and "household management" covered anything that concerned the livelihood of a family, a prince, or a state. The theory of the state's business was called *oeconomia publica*, i.e., public economy or political economy (if the focus of attention was the sovereigns and their well-being, then one spoke of *oeconomia principum*). The remainder was labelled *oeconomia privata.* [2]

All this sturdy wisdom could hardly be accommodated under the roof of moral philosophy and its prudent rules of life or of jurisprudence and its dry judicial expositions. Household management included the art of tilling the soil, mining the ore from the hills, hunting and fishing, keeping the books of large establishments, trading in the towns, collecting taxes for the kingdom. Most economic knowledge was developed and preserved outside the universities. Economic learning, even in the formal sense, was for long the concern of people in practical occupations, and there were few countries in which the separate study of economic subjects had been introduced at the universities by the end of the eighteenth century: Prussia, Sweden, and one or two more.

Elsewhere, an insight into economics flourished well enough without universities. England had the leading economic theorists of the seventeenth and early eighteenth centuries, but none of them was attached to Oxford or Cambridge. Several of the best-known names, such as Thomas Mun and Josiah Child, belonged to men of the East India Company. Charles Devenant did not complete his studies at Oxford, but was later awarded a doctorate in law on rather vague grounds and finished his active life in government service as inspector general of exports and imports. New ground in economic theory was broken at this time in the hectic world of commerce with which the theories were concerned.

This was to a high degree the theory of the expanding middle class, whose strength was growing steadily. The movement known to Adam Smith as mercantilism, which dominated the thinking of the seventeenth century and much of the eighteenth, is largely a theory of trade, of the right way to conduct trade, of the importance of production and consumption to trade, etc. The economists were fond of speaking of the business of the state, but the business of the state was intertwined with private interests. It was still hard to distinguish between state and private in general. The great trading companies augmented the wealth of the few, but officially and according to economic theory they served the interests of the whole country.

In countries with an autocratic government, such as France,

the favored few could feather their nests as employees of the state, and economic theory became the tool of civil servants. In countries where the bourgeoisie was gradually acquiring a more independent political status—such as the Netherlands or England—economic theory could develop a few steps further from the centers of political power.

Posterity has had many hard words for the doctrine of mercantilism, whether it was developed by civil servants, powerful merchants, or university professors. It is portrayed in the literature as a doctrine inimical to freedom, which placed the individual and all his latent initiative under the wardship of the state. But in this respect, the difference between mercantilism and the economic liberalism that followed it is one of degree rather than of kind.

Intellectually and scientifically, mercantilism was for a long time a very advanced system. It had close ties with the scientific revolution of the seventeenth century, and in some respects may be seen as a branch of it. The road from the economic doctrines of Childs and Devenant, or even of the French controller-general Colbert, to the triumphant new science was not a long one. Not that their theories implied a breakthrough in the way that those of Galileo, Kepler, and Newton did. On the economic front, we have to wait at least until Smith to find anything comparable. The world of which economists spoke before Smith was far too complicated and also, in a sense, far too undeveloped to be the subject of a terrestrial mechanics of household management.

It was on a more direct, often purely personal level that revolutionary science and economic theory met. The economists created theories of the balance of trade and noble metals and wealth-creating labor, but as practical men, they were also aware of the need for quite different doctrines, theories on heavenly bodies, the refraction of light, the combination and decomposition of substances. The economists were strict utilitarians; everything that favored welfare and wealth should be utilized. Men such as Petty, Child, and Devenant were sufficiently perspicacious and free from prejudice to understand the utility of the new science. If one knew how nature worked, one would be able to harness it and exploit its riches, all to the greater good of the mother country.

Not until a few decades into the eighteenth century did eco-

nomics became a subject in its own right at any university. The
pioneers were in Prussia (Halle and Frankfurt an der Oder), followed by Hesse (Rinteln), and Sweden (Uppsala). Training in economics did not become a more or less universal feature of the
university curriculum until the late nineteenth century.

All the more importance therefore attaches to the part played
in the development of economics in the seventeenth and eighteenth centuries by a different kind of learned institution—the
academy of science. The idea underlying the academies was that
they should actually counteract the traditionalism of the universities, with their supposed scholastic traditions and devotion to un-
worldly knowledge. "Both the Royal Society and L'Académie des
Sciences represented at the time of their inception the fertile
combination of handicraft spirit and scientific abstraction; from
this was born the experimental research of which they were the
harbingers," says Sten Lindroth.[3] Several of the progenitors of the
Royal Society were themselves versed in the new economic thinking. Particular mention must be made here of William Petty, who
was, of course, one of the great economic theoreticians and also
one of the fathers of modern statistics.

Initially the Royal Society also devoted most of its attention to
practical economic matters. Agricultural questions had an important part to play, and ideas on how new mechanical aids could be
utilized by craftsmen and manufacturers were disseminated by
the Society's publications. Mining technology and practical matters relating to navigation played an even greater part.[4] But after
a few decades, interest cooled noticeably. The Royal Society became an institution in which sophisticated questions of pure science held the stage.

The development of the French l'Académie Royale des
Sciences was in some respects similar. It was closely linked with
the central government, and its promoters and founders saw the
solution of practical problems as one of its main tasks. Here, too,
genuine interest in economic matters soon began to subside. On
the other hand, the Académie had certain surviving obligations
imposed by the state that made it impossible for the Académie
ever to abandon its original direction entirely. It had, for example,
to pronounce on applications for patents. It also published a series
of exhaustive works on industrial and handicraft subjects.

A similar situation prevailed at the academies in Prussia and

St. Petersburg. Utilitarianism had begotten them. It set its stamp on their early years. But then more purely theoretical problems took over.

Population, Enterprise, and the New Knowledge

One may wonder whether the enthusiasm for utilitarianism that inspired those who founded the Swedish Academy of Sciences was not stronger than in any of the countries we have mentioned. In Sweden it was monumental.

Sweden had gone through a period of rapid change after the end of the Age of Empire. This was no less true on the ideological level (and ideology includes the discussion of utilities). Where martial virtues and loyalty to the sovereign had recently been prized above all else, the word was now that the most praiseworthy of all exertions were those on behalf of the national and the private economy. A stream of booklets, brochures, and even newspapers began to appear, proclaiming the new message. Many people bore excited witness to the wealth of England and other European countries.

Pride of place among these has to be accorded Andreas Bachmanson, later ennobled as Nordencrantz. The first part of his book on the secrets of economics and trade, *Arcana oeconomiae et commercii,* had appeared in 1730. It was a demonstratively unpedantic book, for it scarcely contained more Latin than the four words of the title, and it made no reference to learned men and their works. Nordencrantz was himself an unlearned man in the sense of the period, with no academic education whatever. He had been engaged in trade and had lived in England, which had deeply impressed him with its healthy and liberal spirit of enterprise and its lively trade.

One of the principal themes of the young Nordencrantz was that all studies should satisfy the requirement of practicality and usefulness. Neither a young nation nor a young man can start by reading philosophy or classics; economics has to come first. Economics is the origin of all sciences. In their earlier years, even the universities of Oxford, Cambridge, and Leyden devoted their energies to practical matters. Now that the countries in which they are situated had become wealthy, they could afford the luxury of loftier questions. It was, he said, thanks to utilitarian studies that

England and the Netherlands had gained their economic strength. When the stomach has been filled, the mind may need its diversions. But Sweden had not yet got far enough along that road. For that reason he dreamt of a school where *"Oeconomic* and not *Metaphysical studies* shall occupy the place of honor," and he declared optimistically that once one began with economics, the other useful disciplines—physics, mathematics, mechanics, navigation, etc.—would soon follow. When this point had been reached, the nation would become rich and the level of prosperity high, and then, but only then, the way would be open to such adornments of wealth as philosophy, literary criticism, classical studies, even Hebrew and other such subjects.

Nordencrantz never became a member of the Academy. He was once proposed for election, but was not accepted. The reason for this is obscure; perhaps his attitude to knowledge was just too irreverent.

Another eighteenth-century traveller, on the other hand, Jonas Alströmer, was among those instrumental in the founding of the Academy. Alströmer, whose name has come to symbolize the investment in manufactories that was a feature of the Swedish Age of Freedom, gave an address at the end of his presidency in 1745 that was one of the most characteristic expressions of the Academy's utilitarianism. The address was entitled *Sveriges välstånd om det vil* [Sweden's prosperity, if it wishes]. In it he spoke of the impression that England, Holland, Flanders, Brabant, and certain German states made on him in his youth. As a native of the sparsely populated Swedish forests, he was amazed to see people everywhere, "teeming like ants." At the same time, there were few beggars. And the reason? It was the restless spirit of enterprise, the unhindered concentration on useful work: shipping, fishing, manufactories, spinning mills, farming, vineyards, tobacco plantations. . . . Nobody had more iron or copper than Sweden: Holland, Brabant, and Flanders had none. The reason for their dense populations and their full employment was their industriousness. The young Alströmer had his idea: what Sweden lacked was manufactories.

Alströmer's view that the development of manufacturing industry was the strategy on which Sweden would thrive was by no means shared by all the members of the Academy. However, they were unanimous that the prosperity of a country required an abundant population. The assertion that a large population is the alpha and omega of prosperity is a refrain running through the publications of the Academy for many years. Indeed, it is found throughout the huge amount of writing on economics in the 1730s, 1740s, and 1750s. We have to peer into the shadows to discern the opponents to whom these words are repeatedly addressed. There we find those who repeat the old commonsense adage that more people mean more mouths to feed. These sceptics had not been reached by the fundamental mercantilist thesis that there were always twice as many hands as mouths. For according to the mercantilists, it was the work of the many that created prosperity.

This enables us to understand why direct contact with Europe's large populations had both such a disturbing and inspiring effect on Swedish travellers. The problem now was to make Sweden populous, enterprising, and prosperous. One of the means of doing this was through the growth and diffusion of useful knowledge. The Europe seen by the Swedish travellers was a continent seething with new knowledge and revolutionary thinking. The men of the Academy saw a close connection among population, enterprise, and a new utilitarian learning.

In Sweden, therefore, as in many countries, an alliance formed between radical science and radical demands for new industry and new industrial policy. The radical science of this period undoubtedly included mercantilist-inspired economics. It may be difficult for later generations to see any striking similarity between the quickly successful disciplines of experimental physics and chemistry on the one hand, and the type of economics being considered here on the other. But to contemporaries, who had not yet seen what would become of the various imposing projects, things looked different. An eminent physicist such as Anders Celsius also put his faith in economics, and in both sciences he saw the possibility of allying the objective search for the truth with useful application. Celsius was one of those who even in the Uppsala of the 1720s argued for adapting the universities to the needs of a new age. Wholly in the spirit of Nordencrantz, he and others of a like mind declared that a subject such as Latin verse belonged to the ranks of the "less essential." Professorial chairs should rather be established in experimental physics, chemistry, and—economics!

This program soon gained the support of a growing power elite in Sweden. It included Carl Gyllenborg, Anders Johan von Höpken, Carl Gustaf Tessin, and various others. They played a leading role in the political faction known as the party of the Hats (their opponents were the Caps), who put utilitarianism and the radical organization of industry at the head of their manifesto. When the Hats seized political power at the parliament of 1738–1739, they sought in one area after another to hammer home the message of pure utilitarianism. Their assumption of power also came indirectly to play a part in the founding of the Academy. Moreover, Höpken was, of course, one of those on whose initiative the Academy was created; indeed, he came—clearly not altogether accurately—to be honored as its "father and founder."

This does not necessarily mean that the Academy became an institution steeped in the ideology of the Hats. It was the gathering place of all who sought to increase and spread useful learning. The concern for the dissemination of this learning was in its way unique to the Swedish Academy of Sciences and was to some extent a part of the economic ideas prevailing within its ambit. The innumerable findings that were published in the Proceedings were intended—via the clergy and others in authority—to reach the common people. There was not only a concern to produce knowledge that might make the country populous, rich, and powerful. Pains were also taken to see that the knowledge in question reached down to the grass roots. The peasants were to learn new ways of tilling the soil, the miners new ways of handling ore. The activities of the Academy thus came to fit into the long Swedish tradition of popularization.

The Different Kinds of Utility

Economics was *the* useful subject; it was the quintessence of everything that could be called useful. It is thus not surprising that economics played a cardinal role in an extremely utilitarian age.

There were no discordant voices among the adherents of utilitarianism. The things that could be labelled useful were manifold. The reader of the Academy's Proceedings or of any similar series of publications from the Age of Freedom may finish with the impression that any formally rational action—or any zweckrational behavior, if we may anachronistically borrow an expression of Max Weber's—could be counted as useful.

But certain things were excluded, however rationally they were pursued. These included much of the traditional scholarly education. The classical languages might be an aid in some contexts; but to study them in their own right or to learn about the civilizations of antiquity was, as we have already seen, regarded as "less essential." Religion was very useful both for the salvation of the soul and for order in society; but if it conflicted with other vital and, in particular, economic interests, it must not sit in judgment. An immersion in the mysteries of religion was in any case to be deplored.

In view of the strong opposition to the conditions of the Age of Empire, it might easily be expected that military activity would invariably be considered nonuseful. But this was not so. We must not forget that the people who made themselves the political vanguard of the great utilitarian campaign, in other words the high-ups of the Hat party, were also those who worked with gusto for a revanchist war against Russia (which, indeed, they managed, with lamentable results, to bring about). According to them the ultimate goal of all utilitarian activity was to make Sweden strong. Efforts ought, therefore, to be concentrated on this ultimate good.

By strength was meant not only, perhaps not even primarily, military strength; but military strength was nevertheless included. Sweden was to attain a respected position in the world for its trade, its national wealth, and its armies and navy; also for its good morals, its flourishing agriculture, and its manufacturing, and, last but not least, for its science. It was the Swedish state that was to glitter in the eyes of the surrounding world.

The "state" should not be seen here as an abstraction. It was in a very concrete sense the political institutions, particularly the estates of the Riksdag, and the government offices. The state was the top-ranking figures in the government, who were also, during the regime of the Hats, the leading people in economic life.

We must not confuse the utilitarianism of the eighteenth century with that of the twentieth. In the utilitarianism of the twentieth century, which has also dominated the life of the Swedish nation, the welfare of the individual citizen and his standard of living play a decisive role. The typical eighteenth-century utilitarian had no such ideas. On the contrary, they would have struck him as highly objectionable, not to mention ruinous. In his scheme every Swede fulfilled his duty to be useful by working hard and purposefully. Moreover, it was his conviction that work, at least heavy labor, involved such a sacrifice that nobody would do it without being obliged to, and nobody would do more than was necessary for his simple survival. This was sound mercantilist doctrine—and also underpinned by older ideas—and it was expressed by the leading Swedish economist Anders Berch when he said that it was important to keep the majority of the population at a "steady and appropriate level of poverty." Plenty would immediately lead to sloth. But sheer physical destitution was also undesirable. The level of subsistence had to be kept more or less constant, and the state had to do this through strict economic control.

The loyal subject would, therefore, by doing the hard and purposeful work required of him by his masters, do his bit for the good of the nation. Higher on the social ladder, the incentives were different. The man of learning would be driven by his curiosity; the only concern was to ensure that his curiosity was applied to useful questions. The civil servant would be guided by his regulations. In accordance with the principles of pure mercantilism—of which Berch and the leading Hats, but by no means all the other members of the Academy of Sciences, were devotees—the greatest freedom should be given to those engaged in foreign trade. Their activities were, Berch considered, the consummate demonstration of utility in the country. They should be allowed to go about things in their own way, for whatever they did would be of the greatest value to Sweden.

It is evident that a majority of the early members of the Academy gave more or less wholehearted support to the political and economic program forced through by the Hats at the parliament of 1738–1739. They all agreed with the Hats in placing a high premium on useful knowledge and in their dislike of unpractical bookishness.

Mathematics had its uses to the military and in the mechanical arts, as did physics. Chemistry was important to agriculture, mining, handicrafts, and manufacturing; agriculture also benefitted from a knowledge of botany and zoology; the health of the people from medicine. The supporters of utilitarianism added to the list linguistics as an aid to communication; religion for the improvement of morals, especially the most important moral quality of all, namely, diligence; and history, which gave insight into the true nature of Sweden. But the natural sciences provided the focal point of the utilitarians' interest, particularly in the Academy of Sciences.

If utility was such a wide-ranging concept, this soon led, not unreasonably, to the question: but which is the most useful? Where should we start? We need look no further than the presidential addresses to see that there was a repeated lack of agreement on this question. Let us consider a few examples.

We are already acquainted with Jonas Alströmer's address, *Sveriges välstånd om det vil*. Alströmer, the great advocate of manufacturing industry in Sweden, ventured also to depict the relationship among the different utilitarian activities. Manufacturing and handicraft were like the weight in a clock: they made it tick. Agriculture was to sustain these activities. Trade and navigation were necessary to distribute the products.

There was no hesitation here, then, in declaring manufacturing—together with its less glamorous elder brother handicraft—the most useful. Every branch of industry was, it had to be admitted, absolutely essential in its own way. But it was primarily with the aid of manufacturing that the nation could take really impressive strides along the road to prosperity. For according to pure mercantilist theory, it was the refining that was the source of value. The further one got from the raw material, the greater the value. The economic theoreticians made careful calculations showing how much more value was added to a manufactured product, a long way from its natural origin, than to an agricultural product, where the distance was a great deal less. Alströmer knew of these calculations. He was also comfortable with the argument that it was only through an improvement in the manufactories that the increase in population that was so indispensable to national prosperity could be brought about. The manufactories could provide employment for many pairs of hands, but at the same time more food was needed, and this required a livelier agriculture, and so on. According to this line of reasoning, the manufactories formed the mainspring of development.

But Alströmer's view must be regarded as fairly extreme among the members of the Academy. There were other presidents at roughly the same time who expressed different opinions.

In splendid isolation, or almost so, stood the versatile inventor Christopher Polhem in his *Tal öfver den vigtiga frågan: Hvad som vårt kära Fädernesland hafver nu mäst af nöden til sin ständiga förkofring i längden?* [What our dear fatherland needs must have for its continuous and lasting betterment] (1744). Polhem was by this time an old man, which he did not seek to deny. He was against all the foreign influences that he saw Sweden absorbing. The most important thing, he said, was to love one's country and to preserve its "home customs." Learning should be communicated in the native tongue and no other. Those who travel abroad should be closely supervised, and censorship should keep a careful watch on "foreign books." The most Swedish sectors of all were not only agriculture but also mining, Polhem's own particular sphere. Agriculture and mining should be the occupations given priority by our utilitarian society, but manufactories should also be included *to the extent that they were based on Swedish raw materials.* Polhem sang the praises of iron, "the master key to the welfare of the many."

These effusions on behalf of good old Sweden probably struck the Academy as a trifle extreme. However, Polhem was far from alone in taking up cudgels for the country's staple industries. In this respect he seems likely to have been speaking for the majority of the members.

In the same year 1744, Theodor Ankarcrona, a naval officer well versed in the literature of economics, delivered his *Tal om förbindelsen emellan landtbruk, manufacturer, handel och sjöfart* [On the connection between agriculture, manufactories, trade, and navigation], in which he claimed that agriculture was the foundation stone and for that reason had "precedence and preference." The better the state of agriculture, the better the other branches of industry would be.

A year earlier, in his *Tal om et borgerligt samhälles eller ett land och rikes rätta styrka* [On the true strength of a civic society or of a country and kingdom], the impetuous Henrik Jakob Wrede had admittedly emphasized, like the Hat politician he was, that nothing should be left "in its rawness and incompletion" and that everything should be refined as far as possible. But he also thought that the first duty of a legislative assembly was to promote agriculture. To wherever there was food, handicraft would be drawn.

This was a chain of causality running in the opposite direction from that supposed by Alströmer and others. The more purely mercantilist viewpoint of Alströmer was expressed with precision and clarity by the first secretary of the Academy, Jacob Faggot. As secretary, Faggot had to reply to the addresses of the outgoing presidents. When these conflicted too sharply with his own opinions, he did not fail, while remaining duly courteous, to record a protest. We see this at its clearest in his reply to Ankarcrona. He stated here that the branch of industry that "will most quickly bring us gain should be the first tended to," and "then the handicrafts undoubtedly claim first place."

Sten Lindroth asserts that these disagreements are of only academic interest.[5] This is true if we consider them only for their effect on the Academy. The conflicting opinions did not, as far as we can see, create any ill-feeling among its members. The speakers expressed their various opinions with good humor and cooperated successfully in the business of the Academy, filled to a man with utilitarian fervor even when differing on the priority of the various utilities.

This equanimity may have been assisted by the fact that the Academy was not an assembly in which differences of opinion on the importance of agriculture, manufacturing, and trade had to be reflected in decisions of various kinds. The members could content themselves with their rhetoric.

In other contexts conflicting opinions of this kind inevitably had consequences of greater gravity. Such was the case in the political arena. If manufactories were to be given priority, then the government, in particular, would have to put money into them. This is just what happened; the person most favored was Jonas Alströmer. The decision to support his activities was highly controversial and led to a series of political rows. His opponents claimed that support was being given to manufacturing at the expense of agriculture, which in their opinion was far more important to the nation's well-being. This political schism lasted throughout the Era of Liberty and played an important part in the struggle between the Hats and the Caps.

The same clash of opinions may be seen at the universities after the establishment of the first chair in economics at Uppsala University was carried through the parliament of 1738–1739 by the Hats. Anders Berch, the first incumbent of the chair and the only university economist of any standing in the eighteenth century, was a typical and very persuasive spokesman for the extreme Hat viewpoint. He saw manufacturing as the branch of industry that would put the nation's economy on its feet. He also claimed that every good economist must above all else have a command of the branches of knowledge that were of immediate use to manufacturing; only afterwards was there room for the knowledge needed for agriculture.

His views here were opposed by those of Carl Linnaeus. Berch and Linnaeus sat peaceably enough side by side in the Academy of Sciences, where, as we shall see, Berch's contributions were observations most closely relating to the agricultural sphere. At

Jacob Faggot (1699–1777), Director of the Land Survey Board and at one time Secretary of the Academy.

Anders Berch (1711–1774), Sweden's first professor of economics. (Courtesy of Uppsala University.)

the university, where appointments had to be made that involved picking and choosing among experts in different specialties, there arose controversies which, if never vehement, nevertheless dragged on for decades. To Linnaeus, economics meant neither more nor less than agricultural economics. Economics called for a sound knowledge of botany and zoology; anything else was secondary.

In 1759 Linnaeus succeeded, thanks to a donation from a Värmland ironmaster, in establishing a second chair in economics at Uppsala, this time in practical economics. Similar posts, also concerned with practical subjects, had already been set up at Åbo (Turku) and Lund. These became posts for Linnaean disciples. When the utilitarianism of the Age of Freedom had faded away, they increasingly became posts in pure botany; the economics side was forgotten. By this time the professorship that had first been held by Berch had become a position for jurists.[6] That the harmony of the Academy was undisturbed by such differences of opinion was not because these differences were over matters of no significance, but because the fact that different members ranked the utilities differently did not interfere with the Academy's work.

It is now time to outline very briefly how the utilitarian enthusiasm found expression in the many observations presented to the Academy.

Agriculture First

Even if there were incompatible opinions on whether priority in absolute terms ought to be given to manufacturing or to agriculture, there is no doubt that the Academy did come to devote the greater part of its energies to agriculture. This was not only or mainly because its most prolific members—including even advocates of manufacturing such as Faggot—knew more about agriculture and about subjects that were of direct use to agriculture. The supporters of agriculture were assisted by the fact that there was simply a much larger, richer, and more accessible body of experience in agriculture than in anything else. Failings and problems were well known to everyone (it was an age when the vast majority of people, including academics, still lived close to primary industry).

What could be achieved with manufactories? The country had

a few—the leading example was down in Alingsås, Alströmer's. Even if there was agreement that a knowledge of mechanics was valuable in manufacturing, the goal was not technological development. The mercantilist doctrine did not point in that direction. More efficient machines would mean fewer hands occupied, which would negate the whole point of the manufactories, namely, that they would help to boost other industries. Progress towards the main goal—a growing population—would be obstructed. Machines, says Anders Berch, are useful when and only when they do work that neither man nor beast of burden can do.

Berch himself tried to present a clear and comprehensive picture of economic utilities, arranged according to the principles of Linnean classification; his particular instrument in these efforts to inform the public was the economic theater that he set up in Uppsala with the support of leading Hat politicians. To illustrate the work of the manufactories, he exhibited models of the machines that were in use and also specimens of raw materials and products, even typical "weaving faults" (the equivalent of the zoologists' collections of deformities and monstrosities!). But he encountered opposition from the manufactories of whom he had spoken and written so warmly—they did not want to reveal their trade secrets!

Berch's collection of agricultural implements and other exhibits relating to agriculture was by comparison all the more splendid. His collection of ploughs, in particular, has preserved his name for later generations. And it was through the plough that he made his most valuable contributions to the Academy.

The generally wretched quality of Swedish ploughs was a well-known problem. They were light and ineffective. The eighteenth century was the great age of agricultural reform: the reformers in England and France found imitators in Sweden, who also borrowed ideas on new types of ploughs. Anders Berch—who had been taught mechanics and physics by Anders Celsius—dealt with the subject thoroughly. In a well-known essay, which appeared in the Academy's Proceedings for 1759 under the title of "Anmärkningar öfver de svenske plogar" [Notes on the Swedish ploughs], he went in great detail through the function of the plough, the different ploughs to be found in Sweden, and even the names of different parts of the plough in Swedish dialects.

Man first had the idea of a plough when he saw pigs rooting around in the soil, said Berch, with an acknowledgment to Pliny the Elder. The learned allusion was not merely for ornament. In a true mercantilist spirit, the author wished to make the point that what the plough did could not be achieved by manual labor. In other words the plough did not stand in the way of the essential increase in the population. On the contrary, improving it would serve to increase prosperity, for with better ploughs more people could be supported. Berch therefore advised "our good Mechanici" to tackle the problem of improving Swedish ploughs. But they must remember that a simple machine was needed, not one as complex as the works of a clock.

In the meantime there were several different types of ploughs in Sweden. Berch had acquired an excellent grasp of his subject; he had scale models of all the types in his museum of economics in Uppsala. But he was not sure which one was really the best and hence the most promising for technical refinement. He therefore rounded off his essay with an imaginative suggestion. Peasants from all parts of Sweden had, after all, to come to Stockholm to attend the Riksdag. Could not the good members of the Estate of the Peasants bring a plough and a competent ploughman with them? A ploughing contest could be arranged on Gärdet in Stockholm. "The results would then surely show which is the best plough."

Berch was not alone in his interest in ploughs. Many who were more directly concerned with ploughing practice than Berch were calling for technical innovations in this field. Although the ploughing contest never happened, the Academy, with its democratic ambitions, received a big response from the peasantry when it sought reactions to its ideas and its findings.

Other agricultural machines were also discussed from time to time. One machine that did not seem to appeal to the true mercantilists was the sowing machine: it would deprive the sower of his work. But optimistic inventors of varying abilities nevertheless submitted proposals for such a contraption to the Academy. Others suggested new winnowing machines. The results of their efforts were modest, however.

But most of the questions of agricultural economics that were discussed by the Academy concerned not machines but manuring, seed, and livestock. Manuring the soil was a particularly pressing problem. There was a troublesome imbalance in Swedish agriculture between the number of livestock and the amount of land that needed manuring. The peasants preferred to grow cereals and other crops that people could eat rather than to bother with meadows for cattle to graze. The Academy organized information campaigns to reach the ignorant masses via the clergy. Much effort was expended in extending knowledge of suitable meadow plants. This was, of course, the special province of the Linneans. It was as a result of their energies that grass for forage became a more common feature of Swedish agriculture. The main forage plants— clover, lucerne, etc.—were examined; experience from the Continent was considered; and knowledge was disseminated through the Academy's Proceedings, the almanacs, and special leaflets.

Another important side of the Academy's activities consisted in proclaiming the merits of the potato. The Swedish people were now to learn to grow and eat this exotic tuber. Jonas Alströmer, whose interests were not confined to his manufacturing, was the pioneer in this field. The whole Academy was soon involved in the informational activity. Its exertions were backed by more powerful forces—the estates of the Riksdag and the Board of Commerce, which was very active during the Age of Freedom. The great campaign began in the 1740s, after several years of crop failure. The Academy was ready with informative essays on the cultivations at Alingsås and on the more advanced potato culture of the Continent. It was envisaged that the potato would replace cereals in one area after another. Bread was baked with potato flour. In fact the new vegetable found wider popularity in the distilling of aquavit. There are, indeed, unkind observers who say that it was the potato's excellent qualities as a raw material for that purpose that played the vital part in ensuring its success in Sweden and the other countries of what may still be considered "the aquavit belt."

Livestock played a smaller role in the Academy's economic preoccupations than crops. But the subjects of cattle breeding and veterinary medicine were sometimes ventilated. However, the favored creature of the members of the Academy was the bee, so essential both to the propagation of plants and to the people's supply of sweeteners.

Some of the ideas typical of the eighteenth-century mercantilists' anxiety to garner as much of what unharnessed nature offered as possible were the schemes to tame wild animals. In the dense forests, elk could be seen wandering about, doing nothing. Why

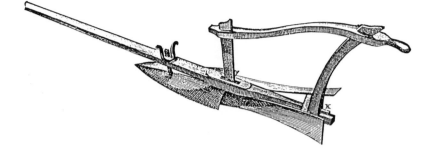

A light iron plough. It was described by Johan Brauner, a leading contemporary expert, in the proceedings for 1749.

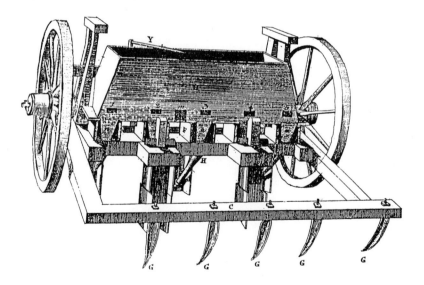

This sowing drill, designed by Carl Johan Cronstedt, was described and illustrated in the proceedings for 1765. The machine could sow five rows at a time.

couldn't the elk be hitched to the plough? He was as strong as an ox and as fast as a horse, declared the most enthusiastic protagonist of this project, Jonas Alströmer's son Claes. It was at Claes' instigation that the Academy offered a premium for every tame elk that could be used for breeding. No claims were received. The grandiose ideas about the elk were among the many unfulfilled hopes arising from the faith in development. But the hopes did not finally die until well into the nineteenth century.

Another exotic project was the attempt to grow mulberry trees in Sweden and establish a native silk production. The scale of the enterprise was large. Plantations were laid out in Skåne; the director of the plantations was to be the talented Linnean pupil Erik Gustaf Lidbeck, professor of natural history at Lund. Not much silk resulted; but tobacco plantations, also a child of the eighteenth century and a concern of the Academy, survived in Skåne until the 1950s. The question of how the forests were to last despite continuous felling also perplexed contemporary economists. The Academy tried to foster thinking about more suitable species and a more active silviculture.

As soon as we leave the sphere of agriculture, the efforts of the Academy at once appear more temporary and haphazard. Attempts were made to help mining over its various difficulties. The art of building houses was another recurrent theme. And throughout the eighteenth century, ideas for improved types of vessels and more economical shipbuilding exercised many keen minds. The members of the Academy also made their contribution to this subject.

It is remarkable how few of the Academy's observations—disregarding those involving agricultural machinery—had any direct connection with mechanics. According to the specification for the professorship in economics that was held by Berch, a knowledge of mechanics and mathematics ranked high among the qualifications of the well-trained economist. The most frequent contributions concerning mechanical devices and contrivances were made by the venerable Polhem, busily active almost until his death in 1751. It is hardly reasonable to expect him to have been as creative and innovative at this time as he was in his youth and his prime. But diligent he certainly remained, and more or less alone in his field among the Academy's members.

However, the leading legacy of the Academy of Sciences in

economics was not in any of the areas we have discussed, but in one where Sweden, or (thinking in terms of modern borders) Sweden and Finland, had a head start on any other country— population statistics. The Academy—mainly in the person of its secretaries Pehr Elvius and, later and even more notably, Pehr Wargentin—was to play an extremely important role in realizing the government's ambitions in this direction.[7]

A number of circumstances were conducive to the production of population statistics. The most important was that every Swede belonged inescapably to the same church: it was easy for the pastors to keep count of people. The web of the Church of Sweden extended all over the vast, thinly populated territory. Moreover, every parish had its definite geographical boundaries; there was never any doubt where a subject belonged. These were the reasons why Sweden succeeded with something that England, more advanced in many ways, found impossibly elusive.

There had been some census-taking before the advent of the Academy. But the great plans were launched in the 1740s. What was needed then was to ascertain the present state of the population resources that were so essential to the prosperity of the kingdom. The interest in population statistics was inspired from without—from England, Holland, and Germany. Anders Berch's *Sätt, att igenom politisk arithmetica utröna länders och rikens hushåldning* [Ways of establishing the economy of countries and nations through political arithmetic] (1746) was an excellent summary of foreign models and also contained a statistics program for Sweden. This was also the subject of earlier essays among the Academy's Proceedings. The method proposed was still that of calculating the total population from knowledge of the numbers of births and deaths. Berch's method was even simpler—the calculation was to be based on lists of persons liable for tax.

However, when the gathering of Swedish population statistics finally got under way, which it did with the establishing of the Office of Tables in November 1748, the figures were not based on calculations but on an actual population count. For this, a smoothly functioning organization was required, with every clergyman supplying up-to-date figures to his rural dean, who then forwarded the information to the central authority.

This was a sizable machine, and it took time for it to be made to work properly. Wargentin played a vital part in this. He also wrote an important essay on the subject, which appeared in the Academy's Proceedings under the title of "Anmärkningar om nyttan af årliga förteckningar på födda och döda i et land" [Notes on the value of annual lists of births and deaths in a country] (1754–1755). It began with a proposition that was typical of the period: "That the greatest strength of a civil society consists in an abundance of good honest citizens is a view on which scarcely anyone now casts doubt." However, it was not so much the rhetorical pronouncements of the indefatigable secretary of the Academy as his practical organizational ability that got population statistics started. A Swedish tradition was born; through the Central Bureau of Statistics and other authorities, it continues in Sweden today.

Conclusion

In his authoritative history of the Academy of Sciences, Sten Lindroth has shown that essays dealing with economic subjects played a large part, possibly the dominant part, in the Academy's early days.[8] It is true that this pattern became less marked, particularly after the Age of Freedom ended in 1772. By then utility was no longer the slogan of the day, and interest in economics was waning. On the other hand, interest in agriculture continued undiminished and even seemed to be increasing. In fact, the Academy acquired a competitor in this field, Patriotiska sällskapet [The Patriotic Society], which was founded in the very year 1772.

But with time the enthusiasm for economic utilitarianism began inevitably to decline. The position of the specialized sciences was strengthened, and the question of what the individual disciplines were to be used for was asked less often. The Swedish Academy of Sciences followed the path already taken by its elder sisters in other countries, in which, however, utilitarian zeal was never so pronounced, nor so persistent.

In certain areas the economic activity of the Academy had permanent consequences. The mere fact of informing and maintaining living links with the public established a popular base for Swedish science in the future. Population statistics continue. The exotic potato survived, whereas the equally exotic tobacco plants and mulberry trees disappeared.

The Academy did not produce, as some of its members hoped, a "Luther of economics." But it did at least contribute to fostering a less dazzling, but larger and economically more literate clergy and civil servant class.

Notes

1. Bengt Hildebrand, *Kungl. Svenska Vetenskapsakademien: Förhistoria, grundläggning och första organisation* (Stockholm, 1939), 371.
2. All kinds of economy were enclosed in the *oeconomia divina*, which meant the whole created world. Cf. Tore Frängsmyr, "Den gudomliga ekonomin: Religion och hushållning i 1700–talets Sverige," *Lychnos* 1971–1972, 217–244. (English summary: The divine economy: On religion and economy in eighteenth-century Sweden).
3. Lindroth, *KVA Historia*, 1:1, 224.
4. Cf. T. Merton, *Social Theory and Social Structure*, 4th ed. (New York, 1968), 680.
5. Lindroth, 233 ff. Cf. also Karl Petander, *De nationalekonomiska åskådningarna i Sverige* (Stockholm, 1921), 115 ff.; Eli F. Heckscher, *Sveriges ekonomiska historia* (Stockholm, 1949), II:2, 858 f.; Börje Hanssen, "Jacob Faggot som ekonomisk författare," *Kungl. lantbruksakademiens tidskrift* 1942, 55 ff.
6. On economic subjects at the Swedish universities, see Sven-Eric Liedman, *Den synliga handen: Anders Berch och ekonomiämnena vid 1700–talets svenska universitet* (Stockholm, 1986).
7. On population statistics, see in addition to Liedman, Karin Johannisson, *Det mätbara samhället: Statistik och samhällsdröm i 1700–talets Europa* (Stockholm, 1988), and, with special reference to Wargentin and the Academy of Sciences, Erland Hofsten, "Pehr Wargentin och grundandet av den svenska befolkningsstatistiken," in *Pehr Wargentin, den svenska statistikens fader: En minnesskrift* (Stockholm, 1983).
8. Lindroth, 226 ff.

ULF SINNERSTAD

Astronomy and the First Observatory

THE changes that took place towards the end of the seventeenth century in astronomy, the branch of learning at the core of what used to be called natural philosophy, were far-reaching ones. Astronomy now became the leading agent of a change in natural philosophy into a practical science calculated to improve the lives of the people. The practical importance of astronomy was obvious. Astronomical observations to determine geographical positions were needed both for mapping continents hitherto unknown to Europeans and for a more exact survey of individual countries and their borders. Traffic over the oceans required accurate navigational fixes for its success and safety, and these in turn demanded in particular a high standard of astronomical timekeeping.

The world's first large observatories in the modern sense were set up in Paris and Greenwich to serve these purposes. But it would be a mistake to see their creation as motivated by immediate practical utility alone. Aristotelian philosophy and the Ptolemaic system had at last begun to loose their grip, at least in those countries that did not feel the full weight of papal authority. Descartes had created a scheme of the universe that had reached the notice of large strata of the population, thanks to a considerable extent to Fontenelle and his best seller *Entretiens sur la pluralité des mondes* (1686). That this was demolished by Newton not long afterwards merely meant that the mechanistic view of the universe that now began to develop had a firm theoretical foundation on which to stand. Astronomy had become a fashionable science, even in the literary salons.

Astronomy at the Swedish Universities

At the Swedish universities, the position of astronomy during the late seventeenth century was insignificant. Neither Lund nor Åbo had a professor in the subject. In Lund, it is true, the professor of mathematics, Anders Spole, had built himself a small observatory in 1672,[1] but it was destroyed four years later during the Danish war and never rebuilt. Lund had to wait until 1813 for its own chair in astronomy.

Following the destruction of his observatory at Lund, Spole was appointed professor of astronomy in Uppsala, with the official title Professor Ptolemaicus. He certainly seems to have had some sympathy for the Copernican system, but as the idea of an earth that moved was contrary to the teaching of the Bible, he settled for a compromise bearing some resemblance to the system of Tycho Brahe. Around the stationary earth in the center of the universe moved the sun, surrounded in its turn by the planets Mercury, Venus, and Mars, and perhaps also Jupiter and Saturn. Whether this was also Spole's personal view is difficult to know. What is certain is that the faculty of theology wielded a lot of power at the university.

When Petrus Elvius was appointed to the chair in astronomy on Spole's death in 1699, the theological pressure at the university had decreased, and it had become less risky to declare a belief in a new cosmology. But Elvius was, like so many others at this time, a Cartesian in his distaste for Newton's theory of gravitation. Its effect through a void over distance seemed nearer to medieval occultism than to the new mechanics. Elvius' successor, Nils Celsius, was another Cartesian, and it was only the next incumbent of the chair, Erik Burman, who finally introduced Newton into the curriculum and gave the first Swedish lectures on Newton's theory. However, Burman was unable to carry out any astronomical observations of importance, since he did not even have a pavilion where he could site his modest instruments. He concentrated instead on meteorological observations and has become known as Sweden's first meteorologist.

Not until the appointment of Anders Celsius, son of Nils Celsius, as professor of astronomy in 1730 did Sweden by fruitful contact with European science lay the foundations of an independent astronomical research tradition. Some years after his appointment, Celsius set off on a long tour of various observatories and centers of learning in Europe. The potential benefit to Swedish astronomy of fresh inspiration and ideas was seen as so great that he was granted nearly five years' leave of absence for his studies.[2]

Even before he left, Celsius had suggested that the Royal Society of Sciences at Uppsala should have an observatory built. It could be financed, as the university librarian, later archbishop, Eric Benzelius had originally proposed, from the profits of publishing almanacs, assuming that the Riksdag would grant the Society this privilege. A request was accordingly made, but the Riksdag turned it down.

When at last Celsius came home from his long sojourn abroad, he did so as an important participant in the famous expedition to Lapland to measure the length of a degree along the meridian, led by Maupertuis. The result of this expedition was that the dispute between the Cartesians and the Newtonians over the shape of the earth was resolved in favor of Newton's theory of the earth as a spheroid flattened at the poles.[3] Celsius now succeeded in 1738 in securing funds to purchase a house in the center of Uppsala. The house was rebuilt to the drawings of the architect Carl Hårleman and was completed three years later. The modest collection of astronomical instruments that belonged to the university was augmented by a number of small instruments that Celsius had acquired during his travels and by gifts and purchases from Maupertuis' expedition. Another purchase was a refractor with a lens ground by the professor of mathematics, the well-known physicist Samuel Klingenstierna, while a transit telescope was ordered from the instrument maker Daniel Ekström. The observatory—apart from Spole's, the first in Sweden—thus had an acceptable stock of instruments, but its location would soon be found unsuitable.

The Stockholm Observatory

The idea that the Academy of Sciences ought to erect an observatory arose early. The plan was in the spirit of the Age of Freedom, with its aim of supporting the nation's progress with practical research. The young academies of Europe, moreover, had set a good example. The Académie Royale des Sciences had built itself an observatory in Paris in 1672, and the Preussische Akademie der Wissenschaften one in Berlin in 1709.

The initiative appears to have come from the secretary of the Academy, Pehr Elvius, son of the Petrus Elvius mentioned earlier and cousin of Anders Celsius. He was well qualified to carry out the project successfully. He had received practical training at the Fortifications Corps in Stockholm under Christopher Polhem, "the father of Swedish mechanics." The theoretical experience had been supplied by Celsius and his colleague Klingenstierna.

Elvius thought the time was ripe when Hårleman was president of the Academy in 1746. Hårleman was the chief commissioner of public works and buildings and an architect, and as such had produced the drawings for rebuilding the observatory in Uppsala. He was best known for his work on the completion of Stockholm Castle and its interior decoration. He was also an influential politician. Sketches and plans were laid before Councillor of the Realm, Carl Gustaf Tessin, who was also a member of the Academy, and a few days later the civic administration in Stockholm received an application for permission to build an astronomical observatory north of Stockholm, on the summit of the ridge known as Brunkebergsåsen.[4]

Presumably Elvius already had plans for financing the observatory when he first made the proposal. It was the old idea that astronomy should be allowed to benefit from a monopoly on the publication of almanacs. The Academy enjoyed better fortune in the Riksdag than the Society of Sciences had done earlier, being granted the license as sole publisher of almanacs as from 1749.[5] In addition, royal letters patent were issued making available to the Academy various building materials, mainly surplus material from the building of Stockholm Castle.

However, a start on building the observatory was not obliged to wait for the revenue from the sale of the 1749 almanac, thanks to an interest-free loan, against the security of this revenue, from Claes Grill, the governor of the Swedish East India Company. And at New Year, 1748, Hårleman was instructed to prepare the drawings, and construction could begin.

During the 1740s the almanac had an annual circulation of about 170,000 copies. It was, in other words, a very popular publi-

The proceedings for 1749 contained a vignette of the new observatory, still under construction.

Astronomical motifs often ornamented the proceedings of the Academy.

cation, which could provide the Academy with a substantial and reliable income and the government with an important mouthpiece for the enlightenment and guidance of the general public (cf. Gunnar Eriksson's article in this volume).

Also associated with the Academy and its almanac is Sweden's changeover to the New Style in the middle of the eighteenth century. Most Catholic countries had already changed from the Julian to the Gregorian calendar in 1582, thereby skipping ten days, whereas the Protestant countries, which did not want to be led by the nose by the pope, changed later. When Denmark and the Protestant states of Germany decided to change to the Gregorian calendar in 1700, the question was hotly debated in Sweden. As shortcomings were seen in both the Julian and the Gregorian calculations, it was decided to work for a more fundamental correction of the calendar, and for the time being not to exclude all ten days at once. The result was that 1700 was not treated as a leap year, and thus Sweden's calendar came to differ from the Julian calendar by one day and from the Gregorian by ten, a particularly inexpedient compromise. The inconvenience of having one's own calendar was underlined when, in preparing the almanac for 1705, Nils Celsius placed Easter a week too late as a result of following the Julian calendar. In 1712 Sweden reverted to the Julian calculation by including two intercalary days in the year.

The question of a calendar reform was raised again in 1750, when the secretary of the Academy, Pehr Wargentin, was commissioned to prepare a statement on the subject.[6] Wargentin pointed out, in a letter that was later to play an important part in guiding the Riksdag's decision, that over the years the spring equinox and the festivals associated with it were gradually moving nearer and nearer to the beginning of the year. He also noted that the public had such faith in the almanac that they arranged their work on the land and in the home according to the almanac rather than the season's weather. For example, he mentioned that farmers in various districts adhered to a time-honored custom of sowing spring seed in the fourth week before Midsummer's Day, which had formerly coincided with the summer solstice. As Midsummer's Day now occurred 14 days earlier in the calendar year, the spring sowing was in effect taking place 14 days later, when the best part of the summer had passed. This point certainly carried weight in the argument to reform the calendar. The reform duly took place in 1753, when eleven days were omitted at the end of February.

Wargentin and Astronomy

Wargentin's contribution to calendar reform was one of his first achievements as secretary of the Academy. His continued efforts in this office over a third of a century meant an enormous amount to the Academy, to the observatory, and to Swedish astronomy in general.

Pehr Wilhelm Wargentin was born in 1717, the son of a clergyman. His father was interested in natural science, and his knowledge was profound enough for him to be proposed as professor of physics at the University of Dorpat in Estonia. He was highly successful in passing on this interest in science to his son. When on one occasion father and son were both watching a total eclipse of the moon, the eleven-year-old boy was so moved by the event that he "from that day forth sought most urgently every opportunity to gain a comprehension of the course of the heavens."[7]

Yet when Wargentin enrolled as a student at Uppsala University in 1735, he planned to train as a pastor, not to study natural science. But at Uppsala he met Olof Hiorter, who was professor of astronomy while Anders Celsius was on leave. Hiorter, like Wargentin, was from Jämtland, and Wargentin quickly became one of his most attentive pupils.

In the autumn of 1737, Celsius returned to Uppsala, filled with new ideas from the observatories of Europe, and now found in Wargentin an enthusiastic assistant in the observations that he was himself able to carry out with the simple instruments he had brought home. The question that Celsius soon put to his eager pupil with the suggestion that it might form a suitable subject for his master's thesis was one of fundamental importance in eighteenth-century astronomy and was also in the contemporary utilitarian spirit of seeking knowledge useful to the country: to determine, by observing occultations of the moons of Jupiter, the exact longitude of the place of observation.

However, the moons of Jupiter were interesting not only because they could serve as a celestial clock that could be read even with a small telescope. They were also of great theoretical interest, since they displayed a planetary system in miniature, in which periods of revolution were stated in hours and days instead of in months and years. The study of the occultations of the moons made it possible to determine their orbital elements and the changes in the orbital elements over time, information that was of the greatest relevance to the problems in the perturbation theory that was being developed during the second half of the eighteenth century.

The first person to draw up ephemerides for the occultations of the moons of Jupiter was the director of the Paris Observatory, Jean Dominique Cassini. His first tables had come out in Rome in 1666 and set the standard for all other works on the subject for over 100 years. He published an improved version in 1693. His son Jacques Cassini, who succeeded him, continued his study of the satellites of Jupiter and prepared new ephemerides that were expected to appear in 1740. At the other leading observatory of the period, Greenwich, these moons and their occultations were a very topical problem that occupied the celebrated James Bradley, discoverer of the phenomenon of aberration.

When Celsius came home from his foreign travels, he brought with him literature on the moons of Jupiter, together with a collection of observations.[8] With the aid of the energetic Wargentin, he began to observe for himself the occultations of the moons from the observatory in Uppsala. As mentioned earlier, these observations need only small instruments, and Celsius therefore considered them a suitable task for the mathematics lecturers of the country's gymnasia, particularly since they might provide a basis for more reliable determinations of the longitude of many towns.

Celsius and Wargentin soon realized that as far as the occultations of Jupiter's moons were concerned, the existing ephemerides were unreliable. Wargentin reported that they were often half an hour out, sometimes even an hour. However, he and Celsius had high hopes of the new tables that Jacques Cassini was publishing in Paris, which would be much more accurate. It was with these tables in mind that Celsius proposed to Wargentin that for his master's thesis he should produce a manual for mathematics lecturers on how the occultations of Jupiter's moons should be observed.[9]

Celsius ordered a copy of Cassini's tables, and Wargentin began work on his thesis by reviewing the history of the discovery of the moons. He followed this with a description of their motions and orbital elements and how their occultations were calculated. But the tables he had ordered were late, and Wargentin was in a hurry because his financial situation was becoming increasingly precarious. Instead of allowing the thesis to be delayed by the nonappearance of Cassini's tables, Wargentin began drawing up his own tables of occultations on the basis of all the observations he had available and those tables that he did possess. At the start of 1741, he showed a draft of the tables to Celsius, who was delighted to find that they were considerably more accurate than those published by Cassini (I). When the tables of the Younger Cassini (II) finally reached Uppsala in the summer, Wargentin's tables turned out to be more accurate than these, too. The difference between the predicted and the observed times of the occultations was found to be only a few minutes. For the innermost satellites, the correspondence was better than a minute.

Towards the end of 1741 the thesis was finished and Wargentin intended to publish it in the *Acta* of the Society of Sciences at Uppsala. At this point, he suffered an unfortunate setback. The manuscript of the tables was stolen, along with his other possessions, during a journey from Uppsala to Stockholm. However, he refused to be discouraged by the loss and worked out new tables with undiminished energy. But once again their publication was delayed, this time by the sudden death of Celsius in the spring of 1744.

They were eventually published in *Acta* in 1746,[10] where for a number of years they were supplemented by new calculations and by comparing calculations with actual observations.[11] Wargentin had a commentary on his thesis published in the Academy's Proceedings in 1748.[12] The work immediately attracted attention abroad and contributed to his rising reputation as the leading expert in this highly important area of practical astronomy.

Wargentin could feel satisfied that his work in science had won recognition both in Sweden and abroad, but he was also alarmed to find that his successes yielded so little material reward. He still had no paid position and considered the prospects of gaining academic promotion at Uppsala to be very limited.

On September 27, 1749, quite unexpectedly, the secretary of the Academy Pehr Elvius died, less than 40 years old, leaving the Academy in grief and consternation. Just five days later, the Academy was ready to elect a successor, and its unanimous choice was Wargentin.

The Academy's astronomical observatory, designed by Carl Hårleman.

It was soon to become apparent that the Academy could not have made a better choice. Wargentin took office and played his part in the most expansive and significant period that Swedish science had ever experienced. As secretary he had a key position in the country's natural science. It was through the academies, not the universities, that research advanced and won new ground. Wargentin was a born administrator and official, but he also always extended a helping hand to anyone less fortunate than himself, whenever the occasion might arise. He was Sweden's link with international research, and above all he was the persevering scientist who, despite his onerous duties, always seemed to have time for his own research in astronomy. He was the leading figure in Swedish astronomy and was in touch with all the well-known astronomers of the world.[13]

The building of the observatory had just started, and the monopoly on the almanac was assured. The future was up to Wargentin. In bringing the building project to a happy conclusion, he had the valuable assistance of the architect, Hårleman. The observatory was as good as ready when Hårleman suddenly died, early in 1753. Neither of the founding fathers of the observatory, Elvius and Hårleman, therefore lived to see its completion. Hårleman's death delayed the opening, but Wargentin and the instrument maker Daniel Ekström moved in during the spring. There they joined the future caretaker Anders Smahl, clerk of the works during the observatory's construction. The ceremonial opening, with King Adolf Fredrik and Queen Lovisa Ulrika as guests of honor, took place on September 20, 1753.

The privilege of selling the almanac had given the Academy steady and dependable earnings, its only income until donors began to appear. The financial situation was far from encouraging, however. The observatory building had cost a lot of money, and the debt to Grill had to be settled as soon as possible. One of Wargentin's first and most urgent tasks was to try to increase the Academy's income. It was a delicate matter to ask to raise the price of the almanac so soon after the monopoly had been granted. Perhaps the Riksdag might, for fear of public reaction, deprive the Academy of its only source of revenue. All Wargentin's quiet diplomacy was needed to push through a 50 percent increase without serious opposition. The press run of 177,000 in 1752, the year before the price was raised, admittedly dropped to 169,000 the next year, but this was still large at a time when Sweden, including

Finland, only had about two million inhabitants. Afterwards, the circulation began to rise again, and the Academy's income almost trebled.

The observatory still had no instruments other than those that had previously belonged to Wargentin and those that the Academy had received as gifts. These were two small telescopes five feet and eight feet long and a small quadrant. With these instruments, Wargentin had first observed from his home and later from the observatory hill outdoors. The earnings from the almanac now appeared secure enough to allow the Academy to set about fitting the observatory with new instruments.[14]

It was taken for granted that the instrument maker Ekström, who had already moved his workshop and smithy into the observatory, would construct the necessary instruments. He was beyond doubt the leading instrument maker in Sweden and was regarded as fully able to match the best abroad. He had studied his trade in London, where he had spent a year and had been on especially good terms with George Graham. He was employed by the Land Survey Office, and also made instruments for Uppsala Observatory and for the Academy, to which he had been elected in 1742. He had been given the title of director of mathematical instrumentation in the realm. The skill of the instrument makers was absolutely essential to the progress of astronomy, but seldom has an instrument maker been held in such esteem as Ekström.

During the autumn of 1754, a portable quadrant of three-foot radius and a four-foot transit instrument were ordered for delivery as soon as possible. These were to be followed by a mural quadrant of eight-foot radius, a reflecting tube, a parallactic instrument, and a levelling instrument of the highest quality. But none of these were ever produced, for Ekström died in the summer of 1755, aged only 44. No other Swedish instrument maker was capable of taking over manufacture, and Wargentin had to turn to England to obtain his instruments.[15] From the well-known firm of Bird in London, Wargentin received a quotation for substantially the same instruments as the firm had previously manufactured for the Greenwich Observatory. These included a mural quadrant and a transit instrument, both of eight feet. However, buying such costly instruments was out of the question, and Wargentin had to content himself with a three-foot, six-inch movable quadrant, which arrived during the summer of 1758. He started using it that autumn and determined the latitude of the observatory to be

59°20′31″3, a value that differs by only an odd second from the true one. This masterpiece of an instrument can still be admired at the present-day Stockholm Observatory in Saltsjöbaden. Two years later a three-foot, six-inch transit instrument was ordered from Bird and a nine-foot refracting tube from Dollond. The famous John Dollond was the first to manufacture achromatic lenses after Klingenstierna had demonstrated theoretically that such a thing was possible. Before this, the clockmaker Petter Ernst of Stockholm had supplied the Academy's observatory with a new pendulum clock that showed seconds and went six months between windings. There were also tubes and smaller instruments, either bought in Sweden or donated by private individuals. The most splendid donation received by the Academy was in 1772 when Gustaf III presented a large collection of astronomical instruments that had belonged to his father, Adolf Fredrik. Many of these instruments were the work of Ekström.

The observatory was now well furnished with different instruments, even if it could not equal the finest observatories on the Continent. The mural quadrant that had been high on the previous shopping list was never purchased. Wargentin himself was satisfied, however, and he particularly appreciated the quadrant and the transit instrument from Bird. In his correspondence with Continental colleagues, he stated that it would not be advisable to incur the cost of excessively expensive instruments in a country with Sweden's climate, where opportunities for observation were limited because in the summer there were three to four months of permanent daylight, and in the winter it was cloudy or so cold that one's hands were not steady enough to handle such delicate equipment. He also wrote that he considered it unnecessary to spend so heavily on astronomy in Sweden when there were already so many observatories in other countries that enjoyed a better climate and were supplied with the most superb instruments and the most skilled observers, "whom I could never hope to emulate, much less to surpass." Wargentin's modesty was sincere. He was an unassuming man who became the central figure in Swedish science and the unquestioned expert in his field in Europe.

From Wargentin's journals it appears that he must have taken advantage of every starry night unless journeys, illness, or other unavoidable obstacles prevented him. Besides that, he had his duties as secretary, which included all domestic and foreign correspondence. He had to edit the Academy's Proceedings for publication every quarter, compile the various kinds of almanac and, for the first ten years of his secretaryship, publish a quarterly essay on "the history of the sciences." In addition to astronomical observations, he agreed to make meteorological ones. From January 1, 1754, he kept daily weather journals that became increasingly detailed as the years went by. The last of his notes was made on December 10, 1783, two days before he died. It also fell to Wargentin's lot to set up and care for the Academy's library. As librarian he obtained many valuable books, thanks to his numerous connections with scientists and institutions abroad. One of the real gems of this collection is Copernicus' first presentation of the heliocentric system: *Commentariolus.* This copy is bound together with one of the second edition of *De Revolutionibus,* printed in Basle in 1566, that had belonged to Hevelius.

Many were the tasks undertaken by Wargentin that had nothing to do with astronomy. Here we will mention only that early in his period as secretary he was asked by the government to help organize the collecting of population statistics. His work here, which soon came to be carried out under a separate authority called the Office of Tables, assumed such importance that he is known as the founder of Swedish population statistics.

The eighteenth century was the age in which astronomers were to reap the fruits of the ground that had been prepared by the pioneers of modern science. All celestial phenomena appeared in a new light, all observations were equally fascinating, and all were equally important pieces in the mosaic of a cosmology that was now developing. Woven into the major questions of the nature of the universe were the fundamental questions of the exact distances within the solar system and the practical questions of the shape of the earth and the precise position of places on its surface. Wargentin gave equally loyal service whether standing alone at his telescope observing all the remarkable happenings in the heavens or organizing Sweden's participation in the great astronomical event of the century: the two transits of Venus.[16]

From Wargentin's journals we see that the number of days of observation in the year was generally between 50 and 80. The first observations recorded in the journal were, naturally enough, of the occultations of the satellites of Jupiter. He continued to observe these for the rest of his life. Accounts of his observations soon began to appear regularly in the Academy's Proceedings, the first in 1752.[17] Here he reported on an eclipse of the moon in December 1749, a bright comet in January 1750, and an occultation by the moon in October 1751. From his continued monitoring of such phenomena, special note may be taken of the bright comet that appeared in 1769 and aroused so much interest that the crowds of comet watchers at his observatory made it difficult for him to carry out his observations. In the Proceedings for 1770, he reported that at least 20 comets had been seen since 1742. "After the large and especially beautiful one that appeared at the beginning of 1744 and the famous Halley's Comet, which reappeared in 1759, we may say that the comet of the year 1769 just passed may be reckoned the finest."[18] However, Halley's Comet was so poorly placed that it could not be viewed from Sweden's latitude, whereas Wargentin was able to track the comet of 1769 for two months. He stated that the length of the tail on September 9 was 50° and calculated that this was equal to five million Swedish miles (i.e., 30 million miles).

Wargentin made particularly important observations of the irregular variable star o Ceti, Mira. He recorded its variations in brightness systematically for 32 years, a unique achievement at this time. He gave an account of the variations in the Academy's Proceedings for 1779,[19] in which he related that the greatest luminosity he observed was on October 30, 1779, when it shone as brightly as Aldebaran. "She was also like that same star, but even more like the planet Mars, as he shone that evening, with a flaring red light." He believed that the variations in luminosity were due to Mira's having spots on its surface similar to those of the sun.

The Academy's Proceedings were written with a patriotic fervor characteristic of the young Academy—in Swedish. However, many accounts were also published in Latin in the *Acta* and *Nova Acta* of the Society of Sciences at Uppsala. Among them we have already mentioned those concerning the occultations of Jupiter. Other accounts published there concern Wargentin's observations of eclipses of the sun and the moon. During the period 1753–1769, Wargentin was able to observe six of the former.[20] Of the 20 eclipses of the moon occurring between 1754 and 1773, he was able to watch seven.[21]

Three particular inventions of the seventeenth century came to play an important part in the flourishing of astronomy over the next 100 years. These were the telescope, the eyepiece micrometer, and the pendulum clock. Accurate determinations of position and time were the foundation on which the practical astronomy of the century was built. Now at last one could grapple not merely with measurement of the distance to the moon but also with that to the sun and hence, via Newtonian mechanics, with determinations of all the distances in the solar system. However, such determinations required measurement from widely spaced points on the earth. The time had arrived for this, too. Long sea voyages were no longer as perilous as before, and the academies of science had become the natural centers for coordinating projects that transcended national boundaries. It was obvious that Sweden, with Wargentin in charge of affairs, would take part in such projects.

A measurement of great practical importance was that of the distance of the moon from the earth or, in other words, the lunar parallax, i.e., the angle subtended on the moon by the radius of the earth. It was essential to know this angle to be able to use the moon's position—relative to the sun and the stars—in the sky for time determinations and thus for determinations of longitude. At that time the solar parallax was the fundamental constant in astronomy and the key to the dimensions of the known universe.

As direct measurement of the solar parallax by determination of the position of the sun is for practical reasons impossible, the solar parallax is measured indirectly by determining the parallax of a planet that approaches close to us. The first attempt to determine the solar parallax from observations of Mars had been made in 1672 under Cassini's direction by means of corresponding observations from Cayenne in French Guiana and from Paris. The ancient value for the solar parallax, 3′, used well into modern times, had thereby been drastically reduced, but the value of 9′.5 now obtained was regarded as uncertain—out by perhaps several seconds of arc.

In Wargentin's first years as Secretary, long before the observatory was completed, an opportunity arose for Sweden to take part in a new French project to determine the parallaxes of the sun and the moon. In 1750 the Académie Royale des Sciences sent the astronomer Nicolas Louis de la Caille to the Cape of Good Hope to chart the southern firmament. He was also instructed to carry out observations to determine the parallax of the moon and at the times of Mars' opposition and Venus' inferior conjunction to observe both planets to obtain a basis for calculating the solar parallax. For determinations of parallax, of course, simultaneous

Pehr Wilhelm Wargentin (1717–1783), the unifying force behind the activities of the Academy during his 34 years as Secretary.

Anders Celsius (1701–1744) was one of the first elected members of the new Academy.

observations from other points were required, as far north as possible and preferably near to the meridian of Cape Town. Berlin and Sweden were suitably placed. Joseph Jérome de Lalande was sent to Berlin, and Joseph-Nicholas De l'Isle, who had been in charge of the St. Petersburg Observatory for some 20 years, planned to visit Sweden. De l'Isle was a member of the Academy and the first of Wargentin's really important foreign correspondents. During his correspondence with Wargentin on preparations for the expedition, however, De l'Isle decided that the Swedish astronomers could take charge of the observations in Sweden themselves.

All the resources of astronomers and instruments in the country were now mustered. Extra appropriations were obtained from the government, both to pay the observers and to acquire instruments. Of the three university astronomers, Mårten Strömer and Jacob Gadolin were to observe from their respective observatories in Uppsala and Åbo, while Nils Schenmark was to travel from Lund to Härnösand, which was believed to lie almost exactly on the meridian passing through Cape Town. The chief astronomical expert on the Swedish-Norwegian border commission, Anders Hellant, was also engaged to carry out observations in Torneå. With the allotted funds, three astronomical clocks were bought, and three micrometers were ordered from Ekström.

Wargentin, who was to be responsible for the bulk of the observations, wasted no time, beginning in May to observe the moon from his home in Stockholm with the aid of a five-foot refractor, and continuing to do so until early in 1752. Five of his observations corresponded with those of La Caille.[23] The result gave a lunar distance varying between 56.502 and 62.694 radii of the earth, agreeing well with the values obtained by Lalande in Berlin. Observations of Mars were carried out by Wargentin in the open on the hill on which the observatory was to be built; he used an eight-foot, six-inch refractor. Here he obtained six corresponding observations around the opposition in September.[24] His result, just over 10″, was later to be found much too high, but it differed little from those of other observers. Wargentin's observations of the moon and of Mars were reported in the transactions of the Académie Royale des Sciences for 1760.[25]

From the Swedish point of view, the parallax measurements were a success. The Swedes had shown that despite their limited resources they could acquit themselves with credit in international projects. The project as a whole, however, had been less successful than hoped. The uncertainty concerning the size of the solar parallax was almost as great as before, and new ways were sought of determining it with more precision.

The attempt to ascertain the solar parallax with the aid of observations of Venus at its inferior conjunction had failed completely. But during the 1760s, there would be two outstanding opportunities to try again. Both in 1761 and in 1769, Venus would pass at its inferior conjunction in front of the solar disc and be seen against it as a little spot. The possibility of using such transits of Venus to measure parallax had been pointed out by James Gregory in 1663 and later elaborated by Edmund Halley in the *Philosophical Transactions* around the turn of the century. The measurements that were needed would in practice be confined to determinations of the times when Venus entered the solar disc and when it left it. If such time determinations were made from places on the earth's surface as far as possible from one another in a north-south direction, then, according to Halley's calculations, it ought to be possible to determine the solar parallax to an accuracy of $\frac{1}{500}$, a degree of precision that La Caille, however, found far too optimistic. The passage of Venus across the disc of the sun takes six or seven hours, and to satisfy Halley's method its ingress and egress would have to be observed from one and the same position. De l'Isle, however, who was the driving force behind the preparations, simplified the method so as to make it necessary only to view the start and the finish of *either* the ingress *or* the egress from the same spot. However, this method called for an exact knowledge of the longitude of the place of observation.

These transits of Venus are infrequent celestial events. They occur in pairs at intervals of eight years, and the next pair does not recur for more than a century. Only two persons are known to have observed an earlier transit of Venus, that of 1639; these were the gifted astronomer Jeremiah Horrox, who died at age 22, and his good friend William Crabtree. Expectations this time were high, and preparations enormous. Never in the history of astronomy had so many people been mobilized, so many expeditions fitted out to travel to so many distant points to observe a scientific phenomenon. No fewer than 120 observers equipped with instruments were sent to 62 different places in those parts of the world where the transit could be observed. Its northerly latitude gave Sweden a strategic position, and the Swedes made good use of this opportunity. With 20 observers at ten different places within the country's borders, Sweden was second in its involvement only to France, which had 31 observers in 16 places. The English initially showed only a cool interest in the transits. But when the French preparations became known, national prestige was suddenly at stake—this was during the Seven Years' War—and the English finally engaged 19 observers and sent expeditions both to the East Indies and Cape Town.

The Academy, which was reluctant on this occasion to ask for money from the government, met the essential costs of the project itself. Wargentin began making preparations in the winter of 1759–1760.[26] He was naturally in charge of the observations in Stockholm. Strömer, Schenmark, and Johan Justander were also in place at their respective universities of Uppsala, Lund, and Åbo. Another important observation point was Kajaneborg (Kajaani) in Finnish Lapland, where Anders Planman was on the scene by the spring, after a long and hazardous month's journey by sledge. Hellant was, as previously mentioned, at Torneå. Other observers had been stationed at Härnösand, Kalmar, Karlskrona, or Landskrona and furnished with the necessary instruments, pendulum clocks, and small refractors.

On June 6, 1761, the weather in Stockholm was virtually perfect. The transit took place early in the morning, and by about half-past three, Venus had entered the solar disc, which it did not leave until just over six hours later.

It was a major event that attracted many spectators to the observatory in Stockholm.[27] In his record of the observations, Wargentin writes: "Present were Her Majesty the Queen, the Crown Prince, a large number of councillors of the realm and foreign envoys, together with perhaps too large a crowd of spectators of all Estates." The latter evidently caused him some inconvenience, for with reference to his assistants who were trying to observe the transit with their own instruments, he comments: "but with the noise of the spectators, those who were in another room of the observatory could only with the greatest difficulty hear the voice of Doctor Gadolin calling the minutes and seconds from the clock." But Wargentin had wished to draw attention to the event and had certainly publicized it himself in advance. On the whole he was quite pleased with both the great public turnout and the observations that were made.

A sensational secondary result of the observations was that for the first time the atmosphere of another planet was observed. Just before Venus entered the solar disc, it was surrounded by a luminous ring, which the astronomers were virtually unanimous in interpreting as the scattering of the sun's light in the Venusian atmosphere. However, there was less agreement with regard to another unexpected phenomenon. This was the "black drop" that formed at both internal contacts, the two occasions when the limb of Venus was tangential to the solar limb from the inside. It had originally been thought that these would be the contacts whose time was most easy to determine. But when Venus came into the sun, it was seen that it did not immediately break contact with the solar limb, but first remained attached to it for a short while by means of a little black bridge, or thick droplet, that did not finally rupture until the planet was already a good way into the solar disc. Wargentin gave the correct explanation: it was mainly a phenomenon of diffraction, and as such, was dependent on the instrument. Different instruments gave different systematic errors in a manner that was impossible to check.

Wargentin gave an account of the Swedish observations in the Proceedings for 1761[28] and made sure that the details were also available to foreign scientists. Preliminary reports appeared, for example, in the *Philosophical Transactions* of the Royal Society for 1761.[29] The material that Wargentin in his turn received from foreign astronomers he forwarded to Planman, appointed professor at Åbo in 1763, who settled for a parallax of 8″.28. James Short in London calculated different weighted mean values from many observations and finally decided on 8″.56, a value that he assumed to lie no more than 0″.1 from the true one. This was not generally accepted. The largest source of error, apart from the "black drop," was probably the uncertainty regarding the longitudes of the places of observation.

After the partial failure with the transit of Venus in 1761, the astronomers of the world began to look forward with even greater interest to the next one, which was due on July 3, 1769. The transit would be visible in its entirety over the Pacific Ocean and, of course, around the North Pole. In western Europe the beginning would be visible in the evening, and in Asia the end would be visible in the morning. Positions in northern Scandinavia and in the South Pacific were therefore of special interest because of their great distance from each other. The transit would last more than 28 minutes longer in the north than in the south, a fact of vital importance to the accuracy of the parallax determination.

The preparations on this occasion were even more thorough than before. The English were now in the vanguard in terms of the number of expeditions mounted. The Swedish investment in the transit was roughly similar to the previous one. This time, however, money was sought from the government for travel expenses, while the Academy was responsible for the cost of the instrumentation. Particular importance was obviously attached to the observations north of the Arctic Circle.

When the transit took place, it was overcast in Pello and Torneå, but Planman had better luck and was able to observe both the beginning and the end of the transit. Venus' entry into the solar disc had also been observed, and its time recorded at many more southerly observation points. Wargentin reported that although the sun was so near the horizon, he and his assistants had been able to make good observations, almost better than expected.

A particularly fortunate coincidence was that a partial eclipse of the sun occurred the next day, June 4. This was very convenient, since it provided a further basis—in addition to the occultations of Jupiter's moons—for calculating the longitude of the widely separated points of observation. In Sweden it could be seen in both Stockholm and Kajaneborg.

Reports on the Swedish observations appeared in the Proceedings for 1769 and in the *Philosophical Transactions.*[30] As on the previous occasion, Planman performed the calculations. He now obtained a somewhat higher value for the solar parallax, 8″.5. Another Swedish astronomer, Anders Lexell, using a different method of calculating devised by Euler, obtained the result 8″.68; this agreed somewhat more closely with foreign calculations, which generally ranged between 8″.55 and 8″.88. The modern value is 8″.80.

Thus the transits of Venus in the 1760s had reduced the uncertainty in the solar parallax from a couple of seconds of arc to a couple of tenths. Not everyone was satisfied, however. The highest expectations had not been fulfilled. The most serious sources of error in the measurements were still thought to be "the black drop" and the imperfect knowledge of the exact longitude of the observation points. The transits had occupied astronomers for over ten years and had aroused a lively interest among the general public, although the phenomenon itself was far less spectacular than comets and eclipses of the sun. The transits of Venus were soon merely an episode in the history of astronomy, but for Sweden they had another, more lasting effect. They had caused the eyes of the learned world to turn to the country, and under Wargentin's leadership Swedish astronomers had shown themselves well able to carry out their allotted part in the first major international scientific research project.

The uncertainty of the measurements of the distance to the sun was due to the unreliability in the geographical determinations of longitude. It is patient work on a small scale that leads to spectacular results on a grander scale. As an astronomer, Wargentin was primarily interested in improving the tables for the occultations of the moons of Jupiter.

The theory of the motions of the four large satellites of Jupiter is one of the most difficult in celestial mechanics. The difficulties lie in the fact that they all perturb each other and are perturbed by the great ellipticity of Jupiter and by the sun and Saturn. The mathematical basis for the direct treatment of such a problem had still not been devised in Wargentin's day. He had to resort to an empirical-statistical method that he steadily improved and refined. For the method to succeed, an intuitive feeling for the effect of the perturbations is needed, even if one cannot actually analyze them mathematically. This intuition was what Wargentin possessed, together with an infinite patience and inexhaustible assiduity in collecting and processing observations. The empirical-statistical method necessitated a continuous watch on the occultations and minor corrections to the tables to make them as exact as possible. Wargentin was engaged in this all his life. The result was tables that remained unsurpassed in their accuracy for almost half a century.[31]

Wargentin's first tables were published, as has already been mentioned, in Latin in the *Acta* of the Society of Sciences at Uppsala in 1746. Comparisons between the tables and actual observations continued to be included in the same journal regularly. Wargentin's tables are found in the most important contemporary volumes of tables and ephemerides. Tables that he had revised were included in a second edition of Lalande's *Tables astronomiques de M. Halley* (Paris, 1759). These provided the basis for the ephemerides in *Connaissance du temps* for the years 1760 and 1761. Improvements to these ephemerides were introduced in 1766 and 1767. Tables revised again by Wargentin appeared in the second edition of Lalande's *Astronomie,* Vol. 1 (Paris, 1771). Lalande used these tables to calculate ephemerides for the four satellites for the years 1775–1785 in *Ephemérides des mouvements*

John Bird's quadrant, purchased in 1758, was the most expensive astronomical instrument in the observatory.

célestes (Paris, 1774). New tables of satellites were also published in *Sammlung astronomischer Tafeln* (Berlin, 1776), and *Astronomisches Jahrbuch* for the years 1781, 1782, and 1783, and also in the *Nautical Almanac* for 1771 and 1779.

When in the 1770s Lagrange took the general law of gravity as the cornerstone of his discussion of the motions of Jupiter's satellites, he used the orbital elements of their motions that Wargentin had calculated. In the third edition of Lalande's *Astronomie,* published in Paris in 1792, Wargentin's tables were replaced by tables prepared by Delambre on the basis of the fundamental work of Lagrange and Laplace in celestial mechanics, but the form in which Wargentin presented his tables was preserved as far as possible. In reliability the new tables scarcely improved on Wargentin's.

The Observatory after Wargentin

Wargentin's life work coincided with the first heyday of Swedish science. When he died in 1783, he left a gap that no one could fill. While his secretarial duties were taken over by the Academy's physicist, Johan Carl Wilcke, Henrik Nicander, Wargentin's astronomical assistant since 1776, became the Academy's astronomer and the head of Stockholm Observatory. Nicander's contribution to astronomy was insignificant. The Academy's Proceedings contain only a few reports on astronomical observations: a number of eclipses of the sun and the moon, a transit of Mercury, and an occultation of Jupiter by the moon. Nicander also succeeded Wargentin in the Office of Tables, and in 1803 he left his post at the Academy to concentrate exclusively on population statistics.

In one respect the Academy really made an effort, and that was the new expedition to Lapland in 1802–1803. The initiative came from Daniel Melanderhjelm, professor in astronomy at Uppsala since 1761 and from 1797 secretary of the Academy, and he got heavy support from Jöns Svanberg, Nicander's successor as astronomer to the Academy.[32] The background was that Maupertuis' results from the expedition in 1736 came to be more and more questioned. It was without doubt that the earth was flattened at the poles, but probably not as much as Maupertuis had found. Lalande, who was a foreign member of the Academy, raised the question in several letters to Melanderhjelm in 1792–1793. The Academy was positive to Melanderhjelm's proposal, and he started to plan the organization. Svanberg went up to Lapland on a reconnoitering tour in the summer of 1799. Through his colleague Delambre, Melanderhjelm could buy instruments from Paris. The government contributed the necessary economic resources. The expedition, led by Svanberg, was successful, and Melanderhjelm gave continuous information to Delambre, who had these reports read at the French Academy. Lalande congratulated the Swedes in a letter: you have solved a problem that has been troubling the scientists for more than 50 years, "l'époque de votre lettre sera consignée dans l'histoire de l'astronomie." Svanberg gave an extensive report in his book of more than 200 pages, which also included a survey of the history of the problem.[33]

He had finally shown that the flattening at the poles were much less than Maupertuis had found, and this result was generally accepted by the scientific world.

The three holders after Nicander of the post of astronomer to the Academy—Jöns Svanberg, Simon Anders Cronstrand, and Nils Haquin Selander—were all also professors at the Corps of Topography and did most of their work in geodesy and cartography. This was not unique for Sweden. During the nineteenth century, astronomy was still a science intended primarily to serve purposes of economic utility, and many European astronomers were occupied with geodesy and cartography. It went without saying that the Academy's astronomer would be engaged in a project as important as the mapping of the country.

It was not until 1870 that geodesy was separated from astronomy and each subject given its own professor at the Academy. The person appointed to the chair in astronomy and also to the directorship of the Academy's observatory was Hugo Gyldén in 1871.

Observatories had recently been set up and equipped at the universities of Uppsala and Lund, while the observatory at Stockholm had virtually stagnated. The most extensive renewal of instruments during the geodetic period of the observatory had consisted in the purchase of a transit circle in the 1820s, which had been repaired in the 1850s. The observatory building, too, was for the most part unchanged since the time of Wargentin. Gyldén had living accommodations built as an extension to the north side of the observatory, enabling the older part to be used exclusively as institutional premises. A refractor with a 19 centimeter aperture and a focal length of 2.5 meters was bought from the renowned instrument firm of Repsold, manufacturer of many of the most important astronomical instruments of the period. For the installation of the refractor, Hårleman's lantern was replaced with an observation tower and an instrument column was built up through the observatory from the foundations. The observatory thus acquired the appearance that it retains, virtually unchanged, today.

The refractor was intended chiefly for various kinds of micrometer measurement, such as of planets and comets, but it also came to be used in attempts to determine the parallax of stars and nebuli—attempts that later astronomers have found far too optimistic.

When Gyldén was called to the professorship, he already had a good job as astronomer at the well-appointed Pulkova Observatory in St. Petersburg. He had also obtained his scientific education from the celebrated Danish astronomer P.A. Hansen, in Gotha. Hansen had introduced him to celestial mechanics, in which he was to do most of his work, a field of activity that he shared at this time with other Swedish astronomers.

Gyldén began with the problems of the motions of the comets and reported his findings in the first volume of the series of publications that he launched in 1879.[34] He later studied perturbation in the planetary motions and tried to solve the problem of the motions of the planetary system using new methods. He had embarked on a massive task that he was never to complete. Of the three planned volumes of his *Traité,* only the first appeared, three years before his death.[35]

Gyldén was also interested in stellar statistics and was a pioneer in the treatment of the basic relationship among the luminosity of the stars, their number, and their average distance from us. He also proposed that the motions of the stars were less irregular than had been hitherto assumed, and that they displayed a common orbital motion around the center of our galaxy. A half-century later, this hypothesis was confirmed in the Oort-Lindblad theory of the differential rotation of the Milky Way.

Unlike his colleagues at the observatories of Uppsala and Lund, Gyldén had no teaching duties. When he was offered a professorship at Göttingen in 1884, he was tempted, but was persuaded to stay in Sweden. In 1888 he was included among the teachers of the new Stockholm Högskola. This marked the start of a collaboration between the Högskola and the observatory in Stockholm that grew stronger as time went by. Gyldén gathered around himself many pupils, both from Sweden and abroad. His connections with foreign scientists were extensive, and from 1889 until his death in 1896, he was the chairman of the *Astronomische Gesellschaft.*

Gyldén was succeeded as the Academy's astronomer and director of the observatory by Karl Bohlin, who held these posts until his retirement in 1927. Like his predecessor, Bohlin devoted himself chiefly to work in the field of perturbation theory. He was particularly interested in the minor planets. The first minor planets had been discovered in 1801, and the number that was known was growing apace. The calculation of individual orbits for each of them for a long time ahead was a time-consuming task because their proximity to the orbit of Jupiter meant that they were all subject to severe perturbations. Bohlin classified the minor planets according to their motion in relation to the motion of Jupiter and was thus able to make his perturbation calculations in groups, which greatly simplified the complex task of working out the orbits.

However, Bohlin did not only devote himself to celestial mechanics but also took part in observations with the transit circle and, in particular, with the refractor. He observed the minor planets both visually and photographically, and also comets, double stars, and variables. In addition he was interested in stellar astronomy and pointed out that the globular clusters are all on one side of the heavens, and that this is explained by the fact that they are distributed around the center of our galaxy. He may have been the first to draw attention to this problem. At any event, the correctness of his supposition was confirmed when methods of determining the distance of these objects were developed.

There was some further renewal of the instrumentation of the observatory during Bohlin's period of office. For the study of the total eclipse of the sun in 1914, which was visible from Norrland, a 25 centimeter reflector was procured, together with a Zeiss spectrograph. The instruments were later used by a number of expeditions to view solar eclipses.

The lack of instruments satisfying modern standards was all too apparent. Almost every branch of astronomy needed larger and more modern instruments for its pursuit than were available at the observatory. This was particularly true for the increasingly important field of astrophysics, which required powerful telescopes for the analysis of starlight. Photographic observations had almost entirely supplanted visual ones, and a dark sky was essential. The new instruments that were required could not be installed at the observatory, which had become encircled in the early years of the century by the spreading city, with its street-lighting and smoky atmosphere. During the 1920s it became increasingly obvious that a new observatory would have to be built on a more suitable site. Bohlin had written to the Academy suggesting this in 1921, but shortage of funds meant that the plans only began to take shape several years later.

In January 1927 the Academy set up a committee to draw up plans and estimates to a new observatory and consider a suitable site for it. The one selected was Saltsjöbaden, 17 kilometers east of Stockholm, in those days a reassuring distance from the light and smoke of the city, but not too far from the Högskola and other important institutions. The finance for the new building project was guaranteed by a donation from the Knut and Alice Wallenberg Foundation and by the city of Stockholm's decision to redeem its right to the old observatory site.

Bohlin was succeeded by Bertil Lindblad, who, although only 32 at the time of his nomination in 1927, had already won considerable notice abroad for his pioneering efforts in several fields. Lindblad was given overall responsibility for the instrumentation of the observatory and made sure that at the opening in 1931 it represented the last word in contemporary technology and was one of the best equipped in Europe. A new epoch, as important as that of Wargentin, had begun.

This era lies outside the scope of the present account. We shall mention here only that Stockholm Observatory became independent of the Academy in 1973, following the end of the monopoly on the almanac. Since then the observatory has been a department of the University of Stockholm. The old observatory building in Stockholm was long used by the department of physical geography at the university, but it was taken over in 1985 by Stiftelsen Observatoriekullen, a new foundation, which is building a museum for the history of science there. It is probably one of the oldest surviving observatory buildings in the world.

Notes

1. N.V.E. Nordenmark, *Astronomins historia i Sverige intill år 1800*, Lychnos-Bibliotek 17:2 (Uppsala, 1959), 101.

2. N.V.E. Nordenmark, *Anders Celsius,* Lychnos-Bibliotek 1 (Uppsala, 1936).
3. Cf. Claude J. Nordmann, "L'expédition de Maupertuis et Celsius en Laponie," *Cashiers d'histoire mondiale* 10 (1966).
4. Lindroth, *KVA Historia* I:1, 378–392; Henrik Alm, "Stockholms observatorium," *Samfundet S:t Eriks Årsbok* 1930, 105–173.
5. Lindroth, *KVA Historia* I:2, 823–875.
6. *Ibid.,* 841 ff.; N.V.E. Nordenmark, *Pehr Wilhelm Wargentin* (Uppsala, 1939), 166 ff.
7. Nordenmark, *Wargentin,* 9; Lindroth, *KVA Historia* I:1, 378–455.
8. Nordenmark, *Celsius,* 189 ff.
9. Bertil Lindblad, "Pehr Wilhelm Wargentins arbeten över Jupitermånarna och modern teori," *Populär astronomisk tidskrift* 1934.
10. *Acta Societatis Regiae Scientiarum Upsaliensis* 1741 (Stockholm, 1746), Part II, 27–66.
11. *Ibid.* 1742 (Stockholm, 1748), Part III, 1–32; *ibid.* 1743 (Stockholm, 1749), Part IV, 18–54; *Nova Acta Regiae Societatis Scientiarum Upsaliensis* (Uppsala, 1775), Vol. II, 249–266; *ibid,* Vol. IV (1784), 129–178.
12. *KVAH* 1748, 167–184.
13. Lindroth, *KVA Historia* I:1, 48 ff.
14. *Ibid.,* 386.
15. Nordenmark, *Wargentin,* 134 ff.
16. See Harry Woolf, *The Transits of Venus* (Princeton, 1959).
17. *KVAH* 1752, 171–179.
18. *Ibid.* 1770, 175–184.
19. *Ibid.* 1779, 249–259.
20. According to the almanac for Stockholm's latitude for the years 1753–1769, there were seven eclipses of the sun during this period.
21. *Nova Acta Regiae Societatis Scientiarum Upsaliensis,* Vol. I (Uppsala, 1773), 156–168.
22. Although Cassini used his measurements to arrive at a figure that lies very close to the modern one, Albert Van Helden has shown that both the calculations and the measurements were very shaky. The measuring instruments in use were not yet up to the task; Van Helden, *Measuring the Universe: Cosmic Dimensions from Aristarchos to Halley* (Chicago and London, 1985).
23. *KVAH* 1758, 71–90.
24. *Ibid.* 1756, 60–72.
25. *Mémoires de Mathématique et de Physique Presentés à l'Academie Royale des Sciences,* 3 (Paris, 1760), 96–116.
26. Lindroth, *KVA Historia* I:1, 399–411.
27. Nordenmark, *Wargentin,* 176 f.; *KVAH* 1761, 152 ff.
28. *KVAH* 1761, 152–166.

29. *Philosophical Transactions,* Vol. LII (1761), 208–210.
30. *KVAH* 1769, 146–175; *Philosophical Transactions,* Vol. LIX (1769), 327–332.
31. Nordenmark, *Wargentin,* 220.
32. Lindroth, *KVA Historia* II, 296–308; information from Sven Widmalm, who is treating the expedition in his forthcoming doctoral dissertation.
33. Jöns Svanberg, *Exposition des opérations faites en Laponie, pour la détermination d'un arc du méridien en 1801, 1802 et 1803* (Stockholm, 1805).
34. *Iakttagelser och undersökningar å Stockholms observatorium,* ed. H. Gyldén, I (Stockholm, 1879).
35. Hugo Gyldén, *Traité analytique des orbites absolues des huit planètes principales,* Vol. 1 (Stockholm, 1893), Vol. 2 (Stockholm, 1908). Gyldén died 1896. Biography by Bertil Lindblad in *KVAÅ* 1939.

GUNNAR ERIKSSON

The Academy in the
Daily Life of Sweden

WE live in a scientific age. To see the truth of this, we need only
consider the extent to which our daily existence is shaped by
technology and information of scientific origin. Electricity, radio,
and television provide obvious and, nowadays, commonplace examples. How long have we had this intimate relationship with
science? Not very long, surely—not with it playing such an inte-
gral part in our lives as it does now. But the road to our present
state has been an interesting one, and we can look back along it
to times when science was slowly but surely claiming the dominant place in our way of life that it has finally come to occupy. In
Sweden's case, many milestones along this road have associations
with our Academy of Sciences.

The Almanac, Health Information, and the Rural Economy

The Academy of Sciences devoted its energies from the outset
to matters of everyday utility—this was its particular characteristic
and distinguished its approach to science from that of the universities, at least in theory. In this general sense, therefore, the Academy was sympathetic in its relationship to the lives of the
people—it was to favor and foster such knowledge as could be
applied to ordinary life. It was thus largely the problems of agriculture, the dominant industry of eighteenth-century Sweden, that
occupied the young Academy's attention and became the subject
of a long series of essays in its Proceedings. For a while it even

published a special series of Economic Proceedings to cater to these practical and down-to-earth interests. These periodic publications were, moreover, written in Swedish to ensure their diffusion beyond the circles of the learned—here, too, the Academy broke with the tradition of the universities, whose output was mostly in Latin, the language of learning.

Yet it was hardly by means of these Proceedings that the Academy made its mark on the lives of the people. Even though they were written in Swedish, they were, after all, read only by a limited group of wealthy landowners and entrepreneurs—for most Swedes, they were difficult to get hold of and surely much too expensive. It was instead through another publication that the existence of the Academy and of science became tangible to the common man: this was the almanac, the indispensable aid of every reasonably alert and industrious peasant.[1] As early as 1747, the Academy was granted the exclusive right to publish the almanac in Sweden, and it first appeared under the Academy's auspices in 1749. Astronomers among the Academy's membership were enlisted for the fairly demanding work involved—first Olof Hiorter, astronomic observer in Uppsala; then Mårten Strömer, professor at Uppsala, Nils Schenmark in Lund, and the Academy's own secretary, Pehr Wilhelm Wargentin (cf. Ulf Sinnerstad's article in this volume).

The ordinary contents of the almanac were established by tradition: there had, of course, to be a calendar section with the days of the week and the name days, the times of sunrise and sunset (the almanac was published in four editions, which was essential in view of Sweden's size, for the latitudes of Lund, Gothenburg, Stockholm, and Turku), the signs of the zodiac, the courses of the planets, and the phases of the moon. Tradition also required weather predictions, worked out with the aid of the 19-year Metonic cycle—these continued into the second half of the nineteenth century. The almanac also contained the dates of the year's fairs and details of postal services and, last but not least, a short essay for the edification of the public at large—in the first edition it was written by Linnaeus and the subject was ale, its uses and abuses.

The number of copies printed was impressive; from 135,000 in the first year, it had risen to 294,000 by 1785. The main channel of distribution was through the country's bookbinders, and it sold particularly well at the fairs that it itself advertised. This meant that there can hardly have been a village in the country that did not have access to at least one copy of the almanac. It was well known to almost everybody. How, then, did it bring science into the lives of the Swedish people?

Naturally astronomy had a part in this process. All the information about sunrises and sunsets, the phases of the moon, the courses and altitudes of the planets was calculated with expert astronomical knowledge and could not have been imparted to the people without the guarantee of accuracy that such knowledge implied. Even before the Academy was granted its privilege, the almanacs were as a rule fairly reliable and not infrequently the work of people trained in astronomy, but it is nevertheless fair to say that the reliability of the information increased under the Academy. Incidentally, these astronomical calendrical data represent one of the earliest forms of scientific information ever offered to a wider public.

But the scientific information did not stop when the user reached the calendar spread for December. Right from the start—and again, this was in keeping with a much older tradition—the Academy's almanac each year contained an essay, sometimes two, in which an expert in some particular field passed on useful knowledge to the many readers. The contents of the essay might vary over a wide range of topics, but its popular nature could virtually be taken for granted. In the early days, it was often a summary of some appropriate article from the Academy's Proceedings, which was thus emphatically brought to the notice of the general population.[2]

The health and sickness of the people were among the subjects that the men of science dealt with most often in the almanac. At first, essays on the illnesses of childhood and their cure occupied a prominent position; Rosén, physician-in-ordinary and a member of the Academy, was a specialist in this field. Hygiene and diet, also taught by a royal physician—Linnaeus' friend Bäck—became the most common themes of the medical essays later in the century. In 1806 the time had come to publicize the newly discovered defense against smallpox epidemics, vaccination, which was being vigorously advocated in Sweden at this time. Aquavit, midwives, and venereal diseases were other subjects that cropped up in the first half of the nineteenth century, after which essays on medical matters became less frequent, although without disappearing completely.

But just as the Swedish people were mostly employed in agriculture, so for a long time most essays in the almanac were concerned with the improvement of agriculture. The cultivation of potatoes and the manuring of fields were the first subjects dealt with—they were central to the rural economy of northern Europe at that period and were an important part of the effort to increase productivity. The planting of less common useful plants, such as hemp and forage grass, was also discussed, as were beekeeping (strikingly often), substitute feed, and many other urgent matters; in fact no aspect of contemporary agriculture passed without comment. In the nineteenth century, the program of the almanac included such subjects as plant diseases, the cultivation of fenland (necessary as population pressure grew), the increasingly exact findings concerning the supplying of nutrients to plants through fertilizers, and, eventually, dairy husbandry. The public can still find a popular essay in each edition of the almanac to this day, although the Academy's exclusive right to publish the almanac ceased in 1972.

In its agricultural essays, the almanac reflects in at least two ways a process of evolution in the Academy and advances in the sciences with which the Academy was concerned. For one thing, the Academy came in time to take an increasingly stringent view of what might fairly be labelled "science"; whereas in the eighteenth century, relatively banal everyday experiences and even decidedly fanciful ideas could be ranged under this heading, rapid progress was made in the nineteenth century towards the stricter standards of today. Second, science itself tended in general to become increasingly theoretical and sophisticated. We shall see that in several respects this development was to affect the way in which the Academy reached the people in their daily lives.

One area in which the Academy was to make a pioneering contribution to the education of the public was in the practical application of mycology. The serious study of fungi first began after the time of Linnaeus, and it became an important feature of Swedish botany as a result of the efforts of Elias Fries (1794–1878) from the second decade of the nineteenth century onwards. After his student years in Lund, Fries was for a long time a junior lecturer there, until he became a professor of botany at Uppsala in 1835. He had become a member of the Academy at an early age. In 1846 he and a colleague reported at the Academy's request on a disease that was ruining the potato crop, during which he touched on the increasingly important problem of fungous diseases, their role, and control.[3]

In his later years, Fries worked, again at the Academy's expense, to promote the use of fungi as food, which seemed a worthwhile task in a country where fungi were not a normal part of the diet. In a large illustrated folio volume, funded by the Academy, he depicted every known form of edible and poisonous fungus[4]— the latter had to be readily identified if the consumption of mushrooms was to become widespread. This work was, of course, far too expensive for general use, but it could still fulfill an educative function: here mycologists and knowledgeable amateurs committed to publicizing the merits of edible fungi could obtain authoritative descriptions of the species concerned, given with the greatest imaginable accuracy. Fries supervised the illustrators with painstaking care. In the early twentieth century, this pioneering work was followed by inexpensive and more easily available books with good illustrations, which have gradually brought fungi into their current place as a small but frequent item of Swedish fare.

The Central Meteorological Office

When the almanac stopped providing weather prognostications, it did not mark the end of the Academy's concern with the weather. On the contrary, it was only after disowning this pre-scientific legacy that the Academy could give the discipline the scientific legitimacy that the times required. The Academy had started recording exact observations, particularly of temperature, at an early date, but its scientific approach to meteorology became of interest to the public only in 1873, when the Central Meteorological Office was established and placed under the Academy's control.[5]

Modern meteorology had secured a footing in Sweden both at Uppsala University and in the Academy of Sciences. At the university the physicist Robert Rubenson, who later became the director of the Office, had introduced daily meteorological measurements at the newly established astronomical observatory in 1865. At his side he had his then assistant, Hugo Hildebrandsson, who reported in an article in the university yearbook for 1870 on his recent study

Swenska
Wetenskaps
ACADEMIENS
Handlingar

För Månaderna

JULIUS, AUGUSTUS och SEPTEMB.

1739.

VOL: I.

För Efterkommande

Tryckt i Stockholm med Academiens egen bekostnad.

Hos JOH. LAUR. HORRN, Kongl. Ant. Arch. Boktr.

The very first volume of *Handlingar* (the proceedings). This series was the pride of the Academy during the eighteenth century.

tour of the meteorological centers of Europe and drew up guidelines for future weather observations in Sweden.[6]

Some of the observation work that had started in Uppsala was later transferred to Stockholm. During the late 1850s, the efforts of the Academy's physicist Erik Edlund had enabled the Academy to set up a series of observation stations all over the country and provide its observers with instructions and manuals—an interesting method of employing the public in the service of science. These activities resulted in annual reports, compiled by Edlund, for the 14 years 1859–1872. Another of the Academy's undertakings was the supervision of the observations, including measurement of the water level, that had originally been started at 20 coastal stations, 18 of them with lighthouses, by the geologist Axel Erdmann in 1848. Edlund's own observations were in some cases situated at those new-fangled centers of communication, the country's telegraph stations. These gave a number of advantages, which Hildebrandsson summarized in his report for 1870:

> "For one thing the time is always accurately known, for another there is always someone present at the station who can carry out the observations at the appointed time without any great inconvenience, and finally one can rely on at least someone on the staff having some knowledge of physics and familiarity with the handling of the instruments."[7]

One of the essential preconditions of the growth of modern meteorology was the realization of an international telegraph network in the mid-nineteenth century. An international exchange of weather bulletins was initiated by France in 1857; now, in a way previously impossible, one could obtain a quick picture of the distribution of air pressure and winds over Europe and its coastal areas each time observations were taken. The data that were reported could in particular be used for issuing gale warnings. This was also one of the aims of Swedish meteorology, which had already joined the international network in the 1860s with daily reports cabled from Stockholm, Härnösand, and Haparanda (at the northern end of the Gulf of Bothnia). In his article Hildebrandsson proposed the preparation of a daily synoptic weather map of the whole of Sweden and furthermore recommended that responsibility for weather-watching be entrusted as soon as possible to a Swedish meteorological institute to be run with state funds. "It is plain to see," argued Hildebrandsson, "that the saving of a few ships or of a single harvest as the result of a warning of gales or impending bad weather pays for this appropriation many times over."[8] With this he had shown, as clearly as anyone could wish, the role that scientific information of this kind could play in the theater of everyday life.

When in December 1871 the Academy, no doubt inspired partly by Hildebrandsson's report, proposed the establishment of a central meteorological office, it also stressed the practical application that meteorology would thus acquire. The main task of the office would be, it was said, to "reveal the laws of the motion of the whole terrestrial atmosphere." With this approach meteorology would be of great practical importance, not least to shipping, in that the scientists would be able "to use the information received by telegraph on the state of the atmosphere in widely separated parts to draw conclusions on the imminence of storms, of which seafarers about to leave port could be warned in sufficient time by a message on the telegraph."[9]

The Central Meteorological Office began its activities in 1873 under the direction of Robert Rubenson, who, like others in leading positions in the Academy, was soon made a professor. Every morning it received reports by wire from nine Swedish and 21 foreign stations. With the aid of this material, a synoptic weather map was drawn every day and distributed in the capital, to the daily press and elsewhere, and a weather report was prepared. By 1912 the flow of incoming telegrams had increased considerably, to 16 from different parts of Sweden and 68 (even more in summer) from abroad. An especially important innovation was the opening in 1906 of a telegraphic link via Denmark with the Faroes and Iceland, areas that had been found to be of great relevance to meteorological forecasting for Sweden. The telegraph was also used, as the Academy had planned, to forward the data that had been collected, and in due course the forecasts themselves, to various ports and to the municipalities that subscribed to this service. In abbreviated form the information was also transmitted by railway telegraph to the major railway stations, where it was posted for the benefit of the general public. On this front, science drew much nearer to the people. Until it was reorganized in 1919, the Office continued as an institute of the Academy.

However immediate the practical applications of meteorology might have appeared, it was some time before its forecasts could become anything like reliable, indeed, before it even started to issue any forecasts at all. The Central Meteorological Office was at first content to publish its lists without predictions. Forecasts were not introduced until 1880, and even then they did not include any precise details on such an important factor as wind velocity: "windy weather" was the general term for all degrees of strong wind. By the 1890s they might warn of "high winds," but not until 1905 were proper gale warnings introduced, after laborious preparatory work by Nils Ekholm. The year 1880 also saw the start of forecasts of precipitation, temperature, and night frost. These activities were expanded from the summer of 1890 with the inauguration of a special afternoon service of information wired from a limited number of stations at 4 P.M. This service was specifically intended for farmers. However, the meteorologist Ekholm could still remark with a sigh in 1912 that the most important forecasts, those of gales, were far too uncertain.[10]

Other meteorological activities had longer-term goals. The annual tables from the observation stations set up by Edlund continued to appear even after he had handed over the work to the Central Meteorological Office. The number of observation points of different kinds tended to rise steadily. From his position at Uppsala Observatory in 1871, Hildebrandsson had organized an observation system covering nights of frost, thunderstorms, and ice conditions, which was taken over by the Central Office in 1882. At this time every Swedish county had one or two agricultural societies, lively associations led by the county governor and other members of the local gentry who were concerned with its agriculture, and these societies worked tirelessly to modernize the countryside. From 1878 onwards they helped to set up hundreds of stations all over Sweden, where precipitation and in some cases also temperature were measured. Meanwhile, a scheme was devised by the meteorological assistant Hugo Hamberg, whereby the Central Meteorological Office—and therefore the Academy of Sciences—extended its scientific activities to a substantial number of private individuals in rural areas, who for a long time observed and recorded meteorological phenomena without remuneration. Few things could have given the people such a clear signal that the age of science was fast arriving. Its instruments, both human and mechanical, could be seen at work everywhere.

In other respects the work of the Swedish meteorologists covered a broad field, with measurement of cloud heights and velocities, classification of cloud formations, and studies of atmospheric circulation and the application of thermodynamics to meteorological phenomena. With the establishment of the Central Meteorological Office, a division of their spheres of activity could be seen as natural, with the Office being in charge of meteorology as it affected the general public and had a practical application, and the universities devoting themselves with fewer distractions to fundamental research and theoretical studies. But theoretical work was not in itself alien to the Central Office, either. Its terms of reference not only mentioned gale warnings, but also stated that it was to process the records of meteorological observations scientifically and publish them, and to monitor the progress of meteorology carefully and seek to contribute to it.

On the other hand, the longer-term research was also relevant to the business of forecasting. Such research revealed structures that were not immediately apparent from day-to-day observation but which underlay the phenomena that could be recorded at those observations. It also illuminated a series of climatic factors that were of significance to agriculture, forestry, or, by no means least, horticulture in different parts of the country—not because of the weather at any specific moment, but because of the long-term effect that, for example, regular mild winters, high annual rainfall, or dry early summers might exert. With the aid of this material, maps could be drawn that divided the country into climatic zones. This was indisputably useful knowledge to local or national politicians who were planning and encouraging a certain type of cultivation or otherwise trying to steer industries that were dependent on vegetable material towards a particular type of production.

Nature Conservancy and the Museum of Natural History

With its combination of exact methodology and capricious subject—the apparently inscrutable weather—meteorology may remind us that in the nineteenth century the study of nature showed two seemingly different faces. Whereas physics and, even more, chemistry symbolize the quantification and theoreticization of natural science, which made it abstract and precise, natural history in its traditional sense of the study of the three realms of nature—stones, plants, and animals—represents the continued concrete treatment of the tasks of research, based on walks in woods and fields and an almost sensuous perception of the objects studied. It is true that biology and geology were also influenced by the attitudes and methods of more exact sciences, biology most visibly in the rapid advances of physiology and geology in increasingly refined mineral analysis, but much larger branches of these sciences retained the element of open-air activity, albeit combined with more accurate descriptive methods and more theoretical systematics. The dealings of the Academy with the sciences also involved knowledge of this kind, and here, too, there were natural points of contact between the Academy and the everyday life of the people.

As Gunnar Broberg describes elsewhere in this book, the increasingly valuable natural history collections of the Academy were brought together in the Swedish Museum of Natural History, which became an autonomous unit under the supervision of the Academy with a handful of assistant curators for different parts of the collections. These collections were on display to the public at fixed times, a service that may be counted as part of the Academy's workaday contact with the population (even though it mostly took place on Sundays, when admission was free). The Academy was fulfilling a significant educational function here, particularly in spreading an understanding of earlier geological epochs and their organisms and also in conveying knowledge of the more recent fauna of tropical and other inaccessible regions. A special whale museum spotlighted the many Arctic and Antarctic expeditions that were a feature of Swedish science in the late nineteenth century. During the twentieth century, the annual number of visitors to the Museum has fluctuated, sometimes quite dramatically, from as low as thirty thousand to more than seventy thousand: the high fares on the Djursholm line were long regarded by the curator as one of the main hindrances to a greater flow of visitors. Be that as it may, the figures were impressive.[11]

But the Museum's collections also represent a different type of contact between the scientific world and the layman. For this contact is not entirely one way, with science giving and the public receiving. The collections have been heavily dependent for their enhancement on the interest and generosity of the public. And thousands of objects have reached the Museum of Natural History as gifts from private individuals, gamekeepers, forest rangers, teachers. The Museum has indeed shown how much it has appreciated this kind of assistance from the public: in its annual reports every donor and donation have been faithfully recorded. They make both touching and amusing reading, as the eye runs over the acknowledgments for crows with deformed feet, albino great tits, elks' legs, and pheasants of doubtful gender, to dwell for a moment on the "heads of two elk which had gored each other to death on Mackmyra."[12] Of course, the public glimpsed here is hardly a representative cross-section of the Swedish population. In 1910, to choose a year at random, the titles of the donors are recorded as baron, chamberlain, master of hounds, forest warden, inspector of fisheries, house surgeon, regimental veterinary officer, prison governor, master builder, engineer, director, taxidermist, treasurer, student, forest ranger, ironmaster, count, lecturer, major—the number of ordinary Misters (Mrs and Miss do not appear at all in this particular year) may be counted on the fingers of one hand. (The examples are from the list of donors to the vertebrate section.)[13] But even with this restriction to the higher or educated social classes, the social implications of an exchange of this kind are significant.

The publishing of the donors' names in the annual reports in the Academy's yearbook, which were not, after all, that widely circulated, was not the only encouragement given to these contacts. In 1906 the popular journal *Fauna och flora* was launched on the initiative of the Museum of Natural History, carrying articles from the Museum's field of activity, i.e., the study of natural history in Sweden and abroad, and reports on the actual museum work. This journal survives today as one of the guarantors of the availability to the public of copious and detailed information on the scientists' everyday lives.

With this knowledge base in natural history, the Academy was able to assume considerable authority with regard to a question that became steadily more important as a result of changes in society around the turn of the century: the need to protect nature in a technical and industrial society. Views on the preservation of nature had been expressed in the 1880s by the celebrated explorer Adolf Erik Nordenskiöld, member of the Academy and professor at the Museum. But they did not meet with any great response until the new century, when the German conservationist and botanist Hugo Conwentz played an important part with his lecture tour of several Swedish towns in 1904.[14] Karl Starbäck, a lecturer and active mycologist, introduced motions in the Riksdag on the subject of nature protection, which led to the act on national parks and the preservation of natural monuments that was passed in 1910. He helped to organize the voluntary nature conservancy movement that found expression in 1908–1909 in the formation of the Swedish Society for the Conservation of Nature, which met at the Academy. From the government itself, the Academy received the task of supervising the national parks and natural monuments, and it set up a standing committee for this purpose. The importance that the Academy attached to this duty may be seen from its secretary's comments on the act:

> "Under the provisions of these acts the Academy of Sciences has from the start of this year (1910) acquired important duties with relation to the introduction and maintenance of nature conservation work in Sweden. The proportions that this question will assume are impossible to foresee, but it is probable that the protection of nature will in the future demand much work and may perhaps become the greatest task of the Academy."[15]

The commitment was indeed to become a far-reaching one, even if, in the spirit of the period, it long remained primarily a matter of protecting individual natural objects, of which a substantial number were trees growing in an unusual manner and other such curiosities. However, the protection of endangered species all over the country or in certain districts quickly became another important ingredient of the work, together with questions concerning the designation of new national parks or the exemption of individual private interests from the strict regulations within their boundaries. The Academy also contributed to nature conservancy work with a special series of scientific publications that included studies of the natural history of the parks.[16]

The Academy's responsibility for the national parks ended in 1952. But it continued to carry out scientific surveys in the parks as far as resources permitted, and it also had a right of appeal in cases of alleged encroachment. In fact we find that the Academy's activity on behalf of Sweden's natural environment has, if anything, intensified since 1952. The special committee remained active and has had a decisive influence on the preservation of certain

Norrland river systems from hydroelectric power projects. The significance of its work has increased with the growth in our awareness of the macrocontexts in which these environmental threats are set. In recent years the Academy has been giving the scientific study of our environment a global dimension with its journal *Ambio.*

Standard Time and Time Signals

On several occasions in the nineteenth century, the almanac notified the public of changes in the national system of coinage, weights, and measures. This marked the connection between the ruling powers of Sweden and the country's leading learned academy: in the matter of the weighable and measurable in society, the state allied itself with the sciences whose greatest successes clearly lay in the spheres of life that were quantifiable. This sector was to have an importance in the daily lives of the Swedish people that was to grow at about the same rate as the influence of science in general. Here we are considering not the gentle influence represented by natural history, but the precision and exactitude that the mathematical and physical wing of science had made the highest norm.

This brings us to probably the most spectacular events ever to have linked the Academy to the everyday life of the general population. It began on a modest scale in 1841, when the almanac began to show the times of sunrise and sunset as mean solar times rather than, as previously, in apparent solar time with the complications that the latter entailed with regard to checking the correct time-keeping of clocks. The almanac for 1842 gave a clear and easy-to-read presentation of the two kinds of solar time, pointing out the advantages of the one now introduced.[17]

The mean solar time that was given applied to each place individually and depended on its particular longitude. The swift expansion of the railway network that took place after the middle of the nineteenth century accentuated the need for a common time for the whole kingdom or perhaps for an even larger geographical unit. The complications became obvious when it came to indicating the exact times of departure and arrival in train time-tables. In Sweden a general railway time was introduced that was synchronized with mean solar time at Gothenburg, the most westerly place of any importance on the railway system. This meant that all passengers to the east had to get used to times that were behind the mean solar time shown on their own watches, but at least there was no risk of their arriving at the station too late. But in Stockholm, in the east, local time was no less than 24 minutes ahead of railway time! In an effort to reduce the confusion, Swedish stations were in 1864 given clocks with double minute hands, one of them keeping local time and the other Gothenburg time. The telegraph service, on the other hand, chose Stockholm mean solar time as its standard. In these circumstances the introduction of national civil time for the whole country soon became a constant topic of discussion in circles ranging from railwaymen to economists. After hearing various expert opinions, the minister of the interior decided on November 7, 1864 to ask the Academy of Sciences for an opinion on the introduction of a national civil time for the whole of Sweden and to suggest the meridian on which it should be calculated.

The Academy assigned this task to its members Nils Haqvin Selander, the Academy's astronomer and permanent deputy secretary (and also the editor of its almanac), and Georg Lindhagen, its deputy astronomer and a couple of years later its permanent secretary. The Academy must have been well prepared: on November 16 of the same year, it delivered its reply. The action of the minister and the Academy did not produce immediate results, but after new moves in the Riksdag, the decision was taken in 1878 to introduce a Swedish civil time for the whole country as from 1879; it would be calculated from the meridian three degrees west of that passing through the observatory managed by the Academy in Stockholm, which was fully in accordance with the Academy's earlier recommendation. (This meridian runs roughly through the middle of the southern half of Sweden, in which most of the country's population is concentrated.)

In terms of the Academy's impact on the daily life of the nation, the introduction in 1894 of a system of time signals transmitted by telegraph from the Stockholm Observatory may well have been equally important.[18] This was also used—at first, perhaps, mainly used—by the navy, but the signals were transmitted over the telegraph system and reached civil customers as well. Thus, the Academy became involved with the fastest modern means of communication in both its meteorological and its astronomical activities.

By 1900 it was time to replace the nationally determined civil time with the Central European Time that had been thrashed out internationally, calculated from the fifteenth meridian east of Greenwich, splitting Sweden roughly down the middle and giving a time difference of one hour from Greenwich time. The previous Swedish civil time actually differed by only 14 seconds from the new international one. But with increasing telegraph and telephone traffic even this apparently insignificant difference was becoming inconvenient. Here, too, the Academy was able to help, at the request of the telegraph service, with its expertise and its time signals.

Weights and Measures

A practical matter in which both astronomers and physicists from the Academy were able to supply assistance was the introduction of the international metric system in Sweden in the 1870s and 1880s.[19] Scientists had in fact been calling for a universal system of simple and exact measurements since the seventeenth century, and the metric system, introduced in France after the Revolution, soon found adherents among scientists in Sweden. The Academy's physicist Jöns Svanberg had used it in his voluminous report on the expedition to measure a degree of the meridian in Tornedalen in 1801–1803. In an opinion dated August 1823, the Academy in the persons of its spokesmen Svanberg and the chemist Berzelius, renowned for his accurate determination of the weights of elements, recommended only that the current Swedish units (ell, foot, etc.) be related to the decimal system, and their ideas were presented in the almanac for 1825. But even this modest proposal had to wait until 1855 for implementation—local interests, who had difficulty seeing any advantage in the new uniformity, counted for more than those of the internationalists, who undoubtedly included the scientists. But the time was soon ripe. Not only science, but also industry, began to demand reform.

Following the recommendation of the International Geodetic Congress in Berlin in 1867, an international commission met in 1870 and 1872, which in its turn no doubt gave impetus to developments in Sweden. The Swedish decision to introduce the metric system was taken by the Riksdag of 1876, after the Academy's members Fabian Wrede (scientist and artillery officer), Erik Edlund (Academy physicist), and Georg Lindhagen (Academy astronomer and permanent secretary) had prepared a report to the government; the decision was implemented by an ordinance in the Academy's eventful year of 1878, when it was decreed that the system should be fully introduced by the end of 1888. In the international work of deciding on a definitive prototype meter, Sweden was involved from the start. When in 1889 the international prototypes had been manufactured in Paris and were allocated to the participating countries by drawing lots, Sweden received prototype meter number 29 and prototype kilogram number 40. They were both brought home to Sweden by the Uppsala physicist Robert Thalén and kept at the Academy in a fireproof safe, one key to which was held by the secretary of the Academy and the other by the Ministry of Finance—an eloquent distribution. Only in 1935 were the prototypes transferred to the Royal Mint and Department of Weights and Measures.[20]

Against the background of these events, the development of the Academy's relations with the public may be seen both as a progression from the concrete to the abstract and as a gradual transfer from a more casual and superficial to a more fundamental and permanent relationship. The concrete advice on agriculture and health gave way to the abstraction of scientifically determined measurement of time and space; agricultural advice and mushroom-picking were things that one could take or leave—but the new methods of recording time and using measures were unavoidable factors in the lives of all Swedish citizens. The difference lay both in the development of science and in changed social circumstances.

Taking the second of these factors first, it is important to remember that Sweden was industrialized late, but very rapidly.[21] The beginning can be seen in the 1850s, when the products of the Swedish forests, particularly sawn timber, became profitable and steadily more sought-after exports, while at the same time the railways began to stretch out over the far-flung but thinly populated country. But the process of industrialization really gathered speed in the 1870s, when the pulp mills expanded rapidly and ironmaking grew into a major industry. The possibility of exploiting the huge ore deposits of the north and the new technology that could transform the energy of Sweden's many rivers into electric power further heightened the intensity of the industrial revolution in the 1890s.

It is therefore no coincidence that the new and far-reaching standardization of time and measures occurred in the 1870s and the 1890s, a standardization appropriate to the mechanized precision and uniform methods of work of the industrial processes. For such a state of affairs, contemporary science was ideally fitted.

It was not merely that this science had itself reached a more nearly perfect degree of precision. On a much wider scale than previously, science had also shown itself applicable to advanced technology and was thus a factor of maximum significance in industrialization. This was patently true of organic chemistry, which was especially exploited by German industry, and of the science of electricity, where the telegraph, the electromagnet, the generator, and electrolysis give some of the most striking examples of how advanced theory could shape a practice of universal importance.

The successes of natural science in both theory and application allowed the number of people engaged in research to rise, as the demand for them grew, and led to a marked strengthening of the authority and general status of natural science (and hence of the sciences as a whole). The Academy of Sciences, with its epithet "Royal," had always been something of an authority, a government department, but it had perhaps never appeared with such an obvious scientific and social aura as at the end of the nineteenth century. The examples just quoted show how it was now able to influence the very nerve fibers of industrial society.

For what the examples show is that daily life itself was colored by the highly esteemed qualities of modern science: abstraction, precision, and clarity.[22] Measurements were no longer taken from the familiar parts of the human body—in feet, ells, and fathoms—but from a mathematically calculated prototype, carefully watched over by the scientists. And time was no longer measured by the eye's assessment of darkness, light, and the height of the sun, but by telegraphic signals from the astronomers. At the same time, measures became uniform from market to market and allowed less and less room for haggling and adjustment, while accurate time measurement enabled uniform punctuality to become an important element in the lives of industrial workers. The remorseless summons of the factory hooter and the uncompromising whistles of departing trains were once and for all subject to the scientific norm.

These things fall into a pattern that applies to the role of

Elias Fries (1794–1878), a botanist who contributed to the popularization of natural history.

science in society as a whole. For it may fairly be said that the end of the nineteenth century and the first years of the twentieth were generally a time when science began to enter the most central functions of social and private life as never before. In other areas, too, the Academy was able, through its members, to take part in this process of integration, even if it never took place with the same direct institutional connection as in the cases of timekeeping and standardization of weights and measures.

In probability calculations, statistics, and higher mathematics, applications now arose that affected such fundamental subjects as security and freedom, life, death, and democracy. Hugo Gyldén, astronomer of the Academy from 1870, and actually an expert on the mathematics of orbital elements in theoretical astronomy, was a member of several of the committees set up in the late nineteenth century to sort out the problems of proposed pension schemes. He was one of the founders of the Thule Life Insurance Company, sat for a while as the chairman of its board, and served as the company's actuary. Anders Lindstedt, a member of the Academy from 1889 and professor for a time at the Royal Institute of Technology, although most of his research was in astronomy, sat on several government insurance commissions from the 1890s onwards, became a supreme administrative court justice in 1909, and was a driving force behind the pioneering Pensions Act of 1913. Carl Charlier, professor of astronomy at Lund from 1897 and a member of the Academy from 1898, could be engaged in 1920, on the strength of his important research in mathematical statistics (actually undertaken to elucidate the structure of the Milky Way), to investigate the possibility of starting a national lottery. In 1898 he had been asked to prepare material to calculate fare zones on the Swedish State Railways. The pure mathematician Ivar Fredholm, professor at Stockholm Högskola and a member of the Academy from 1914, had a respected name in the Swedish insurance world for his practical formula for calculating surrender values, and he held posts in the administration of both national and private insurance schemes. His tutor and predecessor, Gösta Mittag-Leffler, a member of the Academy from 1883, was on the board of an insurance company and also founded the Actuarial Society of Sweden in 1904. Another of his pupils, Edvard Phragmén, who was also professor for a time at the Högskola (and a member of the Academy from 1901), was later director-general of the Insurance Inspectorate and from 1908 managing director of a life insurance company, Allmänna Lifförsäkringsbolaget. He was included for his mathematical abilities on the proportional representation committee in 1902–1903, and on the proportional voting method committee in 1912, and with this played a part in deciding how the voices of the people were to be heard in Swedish politics.

Although Gyldén and the others who have been mentioned exerted an influence in questions that delved deeply into the private life of the individual, none of them did so specifically as a representative of the Academy. Nevertheless, the application of modern knowledge and the Academy's own life were inextricably interwoven. It is not surprising, therefore, that we find among the Academy's scientific publications some that relate to the same fields. Examples of this in the 1890s, typical period in this respect, are the writings of Phragmén on proportional representation, those of the physicist and meteorologist Rubenson on formulae for calculating the capital value of certain forms of life insurance policy, and those of Gustaf Eneström, perhaps best known as a historian of mathematics and a bibliographer, on pension funds.[23] Maybe the fact that the almanacs of the early twentieth century remarkably often contained articles of a social content relating to these spheres of interest also to some extent reflects the real connection between abstract science and everyday society.

Although science has since continued unremittingly to increase its influence on our daily lives and its integration in our culture, the part played by the Academy in this process reached its peak during the decades around the turn of the century. This was due to a combination of circumstances, all of them more or less arising from the type of institution that the Academy was and is.

When during the last century, science entered, both in Sweden and internationally, the phase of enormous expansion that is continuing to this day, the Academy of Sciences also grew in size.[24] Its astronomical and physical institutes were given new tasks and instruments. The Museum of Natural History was extended, acquiring in addition to the existing botanical, entomological, mineralogical, invertebrate, and vertebrate sections a paleozoological (1864), a paleobotanical (1885), and an ethnographical one (1900), each with a curator of professorial status and authority. The curatorship in botany, which had been linked with the Bergius chair donated to the Academy in the eighteenth century, was detached from this professorship in 1904, creating an additional scientific position. The newly established Central Meteorological Office, as we have seen, was placed under the Academy in 1872; the Kristineberg Marine Biological Station opened in 1877; and in 1905 the Academy's first Nobel institute, in physical chemistry, was set up. A purely material expansion took place when at the start of the First World War, the splendid building of the Academy and the huge Museum of Natural History were completed at Frescati in a newly urbanized area of Stockholm, which gained the distinctive nickname *Vetenskapsstaden* (Science City).

But the growth of the Academy could not continue to keep pace with that of science in general. Both the number of Swedish and international scientific bodies and the number of new disciplines, arising from a specialization within old fields or from hybridization between different disciplines, grew immeasurably faster than the Academy itself. And in these circumstances, the applied science fed to the public by the Academy inevitably became a decreasing fraction of their total information intake from scientific bodies.

Another factor that was bound to affect the Academy's external relations was the general ideology of most scientists—or at least of the most influential ones. It is, of course, well known that the pioneers of the Academy were to a large extent utilitarians, and even if the eighteenth-century concept of utility was not the same as our own, material utility, and with it the application of knowledge in economic contexts, was clearly central to the Academy throughout the eighteenth century. To this was linked a broad concept of science: all kinds of knowledge, even of a very practical nature, such as knowledge of manuring and tillage, might by many people be called science, with no objection from any quarter.

This changed rapidly during the nineteenth century. Berzelius' secretaryship, in particular, saw the quality of science as science becoming a primary consideration. Science became an essentially theoretical and sophisticated search for knowledge—even if, of course, the empirical collection of material and experimentation did not lose status—and practical application had to take place as and when the progress of science towards pure truth allowed the opportunity to arise.

At the same time, a change—wholly in keeping with this new view of science—took place, as we have seen, in the actual character of the knowledge that in various ways reached the public from the Academy. The nature of the knowledge shifted from simple agricultural hints and rules of health to norms for measures, weights, and time indication. We may see the meteorological reports, which retained as much imprecision as the old practical hints, but were nevertheless based on data collected and processed with a rapidly increasing degree of accuracy and frequency, as an intermediate stage between the old and the new. The message of the Academy to the general public was acquiring a steadily stricter theoretical foundation.

The course of events may also be seen as a change in the Academy's involvement with technology in the broad sense, a change that is at least partly parallelled by the change in the forms of technology that have been dominant in society at different periods. In the eighteenth and early nineteenth centuries, agricultural technology undoubtedly occupied this position; with the breakthrough of industrialism, technical interest in agriculture in no way declined, but it accounted for only a small part of a technological expansion that occurred almost entirely in manufacturing and communications. In 1812 a separate academy, the Royal Academy of Agriculture, was set up in Sweden. With this the Academy's involvement with this still fundamental area of technology may be considered to have come to an end, or least to have been sharply reduced. No other area of technology was ever to find its real home in the domain of the Academy, even though to this day the organization of the Academy has always allowed for a special technology section, albeit under a variety of different names. Therefore, although there has always been a significant number of engineers and industrialists within the ranks of the Academy, they have seldom influenced it in this respect. It may be argued that they made their views known when voting on the Nobel Prize selections before the First World War and that they exerted their influence when the Swedish inventor Gustaf Dalén won the prize for physics in 1912.[25] But generally the Academy has been the supporter of pure research—fully in the spirit that has characterized the scientific community since the first great impact of science on society: the late nineteenth century.[26]

As we have seen, the closest dealings with the general public have taken place in matters that concern the wildlife of Sweden—both through the Academy's involvement in nature conservancy and environmental research and, for as long as the Museum of Natural History was under the Academy, by keeping this Museum open to the public and receiving on behalf of the Museum donations from the public in the form of plants, animals, and other natural objects. This, too, can in essence be interpreted as the expression of a "purely" scientific interest: a recurrent motive for the Academy's efforts to protect threatened species has been the need for empirical material of this kind for research. The objects preserved in its Museum of Natural History also served research purposes. It was quite consistent with this to display them to the public as a means of spreading information about the research.

The whole complex of questions relating to the Academy's relationship with the public is, of course, to be seen in the context of the research policies not only of the Academy, but also of the ordinary political authorities of the country. The Royal Academy of Agriculture was founded on the initiative of Crown Prince Carl Johan (i.e., Marshal Jean-Baptiste Bernadotte who had been invited from France). This political act gave the Academy of Sciences a fairly direct reason to drop agriculture more or less immediately from its range of activities. The Central Meteorological Office was placed under the Academy's direction by a decision of the Riksdag after a draft bill from the government; whether this was preceded by active lobbying from the Academy remains to be investigated. When the Geological Survey of Sweden was set up about ten years earlier, this was apparently strongly recommended by the Academy, which, nevertheless, was not put in charge of the project. In some cases there appears to be an element of chance in the way in which new scientific bodies, created in response to the needs of society, were directed and in the role the Academy actually played in the eyes of the government. The National Hydrographic Office was set up in 1908, as an obvious parallel to the Central Meteorological Office, but strangely enough it was not placed under the supervision of the Academy, but given an autonomous position under the Ministry of Agriculture.[27] And when in 1919, it was amalgamated with the Central Meteorological Office, the new unit was entirely independent of the Academy. The general consequence of these developments was that the Academy's modest expansion occurred mainly in the field of basic research.

The same pattern may be seen in the Academy's publishing activities. The emphasis of most of its own scientific literature after the turn of the century was on fundamental research, and the international languages, after the Second World War nearly always English, also took over. And in the 1920s, the secretary of the Academy was able to state unambiguously at an annual ceremony that the primary concern of the Academy was the publication of scientific works.[28]

It is, nevertheless, a mark of the growing importance of science in our culture that this move towards basic research has not reduced the Academy's influence in matters that greatly affect the daily life of every member of the Swedish population.

Notes

1. Lindroth, *KVA Historia* I:2, 823–867; II, 539–570.
2. Greta Philip, " 'Prognostica' i Kungl. Vetenskapsakademiens små almanackor för åren 1749–1954," *KVAÅ* 1954, 343 ff., and [G.E. Klemming & G. Eneström], *Svenska almanackor och kalendarier 1749–1879* (Stockholm, 1879), both works with detailed bibliographies.
3. Elias Fries och P.F. Wahlberg, "Utlåtande öfver potates-farsoten," *Landtbruksakad. handl.* 6 (1846), 123–131.
4. Elias Fries, *Sveriges ätliga och giftiga svampar* (Stockholm, 1860–1866).
5. Nils Ekholm, "Meteorologiska centralanstalten," *Nordisk Familjebok*, 2nd ed., vol. 18 (Stockholm, 1913), col. 270–276.
6. H.H. Hildebrandsson, "Om organisationen af den meteorologiska verksamheten i utlandet samt förslag till dess ordnande i Sverige: Reseberättelser," *Uppsala universitets årsskrift* 1870.
7. Hildebrandsson, 48.
8. *Ibid.*, 52.
9. Ekholm, *loc. cit.*
10. *Ibid.*
11. Information from the annual reports of Riksmuseum i *KVAÖ* and then from *KVAÅ*.
12. *KVAÅ* 1909, 194.
13. *KVAÅ* 1910, 193–197.
14. On the nature conservancy movement, see D. Haraldson, *Skydda vår natur! Svenska naturskyddsföreningens framväxt och tidiga utveckling*, Bibliotheca historica Lundensis, 63 (Lund, 1987).
15. *KVAÅ* 1910, 96.
16. *Kungl. Svenska Vetenskapsakademiens avhandlingar i naturskyddsärenden*, 1938–.

17. See J.M. Ramberg, "Om införandet av borgerlig tid i Sverige och andra länder," *Populär astronomisk tidsskrift* 36 (1955), 39 (1958).

18. Ramberg, 18 f.

19. Se Bernhard Hasselberg, "Om det metriska mått- och viktsystemets uppkomst och utveckling," *KVAA* 1908, and S.O. Jansson, *Mått, mål och vikt 1850–1950: Kungliga Myntet 1850–1950* (Stockholm, 1950).

20. Hasselberg, *KVAA* 1908, 213 and note 1. (Cf. Sven Widmalm, "Mått, mål och vikt," *Svensk idéhistorisk läsebok* (forthcoming).

21. *Cf.* Gunnar Eriksson, *Kartläggarna: Naturvetenskapens tillväxt och tillämpningar i det industriella genombrottets Sverige, 1870–1914* (Umeå, 1978), 151 ff.

22. For the following, see Eriksson, *loc. cit.*

23. E. Phragmén in *KVAÖ* 1894, 1896; R. Rubenson in *KVAÖ* 1891; G. Eneström in *KVAÖ* 1895–1896.

24. For a survey of the growth of Swedish science, see Eriksson *Kartläggarna*. Its English summary may also be found in the essay "The Growth and Application of Science in Sweden in the Early Industrial Era," *Tekniska Museet Symposia: Technology and its Impact on Society* (Symposium No 1; Stockholm, 1979), 101–106.

25. Se Elisabeth Crawford, *The Beginnings of the Nobel Institution: The Science Prizes, 1901–1915* (Cambridge, 1984), 166.

26. *Cf.* Gunnar Eriksson, "Motiveringar för naturvetenskap," *Lychnos* 1971–1972,. 164 ff. (English summary, 169 f.).

27. Eriksson, *Kartläggarna,* 45.

28. See, for example, introductions to the secretary's annual reports, *KVAA* 1920, 1921, 1922.

SVERKER SÖRLIN

Scientific Travel—
The Linnean Tradition

THE founding of the Royal Swedish Academy of Sciences came at a time when the empirical method of acquiring new knowledge had been widely accepted as the most fruitful. Francis Bacon's manifesto of experimental science, *Novum organum scientiarum* (1620), was now over 100 years old, and the Academy's precursors, the Royal Society, l'Académie Royale des Sciences, and the Prussian Königliche Akademie der Wissenschaften in Berlin, had pointed the way by prioritizing experimental research. Experiment and systematic observation were the roads to insight into the mysteries of nature and the universe, roads that had stood the test of time and had been broadened and improved.

During the Academy's early development, there still remained great lacunae in the collected knowledge about the life of the planet itself and its appearance. We can select the year of the birth of the Royal Society, 1660, as our point of departure. Thanks to the discoveries of the great explorers of the Renaissance, from Marco Polo to Abel Tasman of Holland (who reached New Zealand in 1642), the grossly inaccurate world map had been adjusted in important ways. America had been discovered, and its southern tip rounded, and colonization had begun. No one believed that the Indian Ocean was a continental sea, as many had earlier, referring to Ptolemy's *Geographia* (which had been recovered during the fifteenth century).

But the existence of the Southern Continent was still a moot point. *Terra Australis* had been sighted, possibly as early as the beginning of the seventeenth century by Spaniards and Portuguese, and definitely by the time Dutchmen, such as Tasman and Willem Hesselsz de Vlaamingh, charted the contours of the continent later in the century.[1] However, it was still not known whether the piece of land that had been sighted and charted was actually the Southern Continent. Nor had any clear picture emerged of the thousands of islands dotting the masses of water making up the Pacific Ocean. For example, the exact location of Tahiti, later so renowned, did not become known until 1767. Uncertainty also prevailed as to what lay hidden in the interiors of most continents. The giant land mass of Africa had only been nibbled at the edges by trading stations; from them, footpaths led a short way into the areas where unscrupulous Europeans fetched slaves. This hardly served the quest for geographical facts. Little was known of the boundless expanses of Siberia and inner Asia—only after the Siberian Expedition led by Count Orlov of the St. Petersburg Academy in 1768–1774 was the thickest darkness illuminated. As far as Tibet and the other Asian mountain countries are concerned, they did not begin to come into focus until the "Trans-Himalaya" expeditions of Academy member Sven Hedin.

This situation concerned practically all continents outside Europe. Northern Canada was to a greater degree *Terra Incognita*—the contours of the Hudson Bay area were first sketched during the final decades of the seventeenth century.[2] The jungles of South America were represented on maps as a great white blank, until Alexander von Humboldt and his French assistant Aimé Bompland spent several productive years there around the middle of the nineteenth century. Even Europe itself had many places that from a scientific standpoint were unsatisfactorily known. Well into the sixteenth century, the northern regions of Scandinavia could still be portrayed as a foggy, amorphous shape, rich in fantastic features—men with the heads of animals, mystical creatures, and other *mirabilia* and *miracula* of nature.[3] When the German philologist Johannes Schefferus, working in Uppsala, published his *Lapponia* in 1673, it became a scientific best-seller—even though Schefferus had never set foot in the areas he described![4] It is not surprising then that in 1733 the satirist Jonathan Swift could still lampoon the rich imagination and lack of any real information offered by geography:

> So geographers in Afric maps
> With savage pictures fill their gaps,

And o'er unhabitable downs
Place elephants for want of towns.[5]

This want of facts was both troubling and challenging to science. Bacon himself seems to have considered travel a worthy and valuable way of accumulating knowledge; one need only peruse the title page of his *Instauratio Magna,* or read his essay *Of Travel.*[6] And the eighteenth century's scientific travelers set out fully receptive to new facts, convinced that they were fulfilling the Baconian program. They sallied forth, as Barbara Maria Stafford says in her great work on the travelogues of the Enlightenment and Romantic periods, with a 'curiosité dévorante': "an openness to everything, and an intellectual boldness—in the best tradition of the Baconians, the empiricists, and the progenitors of the *Encyclopédie*—brand these published ventures."[7] Swedish scholars set out in the same spirit, the majority scientists and disciples of Carl Linnaeus, enjoying the support of the Academy of Sciences in Stockholm, to whom they reported their results.

Despite the growing appreciation of the value of scientific travel, the great European academies, unlike their Swedish counterpart, did not prioritize such journeys. True, both the French and British academies sent out geodesic and astronomic expeditions, as well as supporting natural history travels.[8] Already in 1666 the Royal Society had issued directives for travelers concerning the proper manner of collecting facts and documenting their discoveries.[9] Yet the results were not impressive. Of the 5,336 essays published by the Royal Society between 1665 and 1848, only 77 deal with geography.[10] While this figure can be misleading (as the results of research trips were published under several different subject headings), it still indicates that the Royal Society's interest in movement in space was not overwhelming. The St. Petersburg Academy was much more interested, perhaps because the discoveries waiting to be made lay in the vast expanses of its nation's own territory.[11] The same conditions prevailed to a certain degree for the Swedish Academy of Sciences, which supported research carried out in the lesser-known northerly reaches of Sweden.

This difference in approach between the Swedish Academy of Science and its British and French counterparts is worth closer attention. At first glance it may seem puzzling that academies of science in London, Paris, and Amsterdam did not take a more

The much-travelled naturalists Sir Joseph Banks and Daniel Solander. Both were elected to the Academy as foreign members in 1773. This painting by William Parry (1775) shows them with the young man Omai, from the Society Islands. (Copy painted for the Academy in 1966 by W.J.S. Brown. The original is in Parham House, Sussex, England.)

active role in the exploration of unknown regions. However, when British, French, or Dutch ships set sail for distant shores, they had other motives in mind (even if these motives could sometimes coincide with purely scientific ones); their primary goals were economic and military-strategic conquest. Beginning in the sixteenth century, the Spanish and Portuguese motivated their respective engagements in searching for the Southern Continent with the belief that colossal riches awaited them there. The Dutch and English had similar inducements. In the 1740s Englishman John Campbell wrote that the British should join the exploration of *Terra Australis* and increase trade in the area, as it would eventually make them a "great, wealthy, powerful and happy people."[12] When James Cook embarked on his three famous voyages between 1768 and 1779, the Royal Navy, typically enough, provided both the ships and the financing. The Royal Society had taken the initiative for the first journey, the goal of which was to reach a point in the Pacific Ocean from which the transit of Venus over the sun could be better observed; but it would never have been able to carry out the expedition on its own.

The scientific aspect of that famous first voyage, which was indeed significant, was the result of the spiritual and financial support of the young Joseph Banks. It was Banks who made sure that accompanying Cook were a skillful botanist (Daniel Solander, a disciple of Linnaeus), a scientific draftsman (the Scot Sidney Parkinson and his assistant Herman Diedrich Spöring, from Turku in Swedish Finland), and a portable scientific library, naturally including the collected works of Linnaeus.[13] Ties to the Royal Society were also in evidence, as Banks had just recently been made a member. But that the trip should have a natural history profile was for the most part Banks' idea. And it was not taken for granted that an ambitious scientific project had any place in such expeditions; when Banks increased his demands before the second trip, craving more space on board for himself and his assistants, the Admiralty simply put a stop to his work. The less pretentious German researcher Johann Reinhold Forster was given the job instead, and he brought along his son Georg, a draftsman.[14] Forster's main objective was to determine whether the Southern Continent did exist; science was thus allowed to come along for the ride, but its presence was not worth too much sacrifice. Nevertheless, the scientific results of Cook's journeys were estimable, not because of the efforts of the Admiralty or the Royal Society, but rather thanks to Banks' devoted generosity in defiance of the short-sightedness of these august institutions.

Dutch expeditions to the Southern Hemisphere had entirely commercial aims. The driving force behind these trips was the Dutch East India Company, and their findings were not treated as new discoveries meant to be shared with the scholarly world; they were closely guarded trade secrets.[15] And yet the gardens and libraries of Holland were enhanced by the findings, and botanical institutions in Amsterdam and Leyden were soon among the most magnificent in the world. Such was also the case in England, where one can trace a direct connection between the building of an Empire and the construction of botanical gardens, such as those at Kew.[16] Again, this seems to have resulted more from the dedication of devoted enthusiasts than from any primary goal for undertaking the expeditions. One such collector was Nicolaas Witsen, a skilled diplomat and coordinator of research trips, who was also a friend of Leibniz, an admirer of the cultures of the East, a philosopher, and an indefatigable promoter of the seafaring interests of his native land. New shipments of scientific curios arrived regularly for Witsen's private collection, among them the first marsupials, soon to become the most popular symbol of the Southern Continent.[17] George Clifford, director of the East India Company, created an exotic garden in Hartecamp, rich both in species and interest; it became the object of wider scientific discussion after Linnaeus, employed by Clifford, published his magnificently illustrated *Hortus Cliffortianus* in 1737.[18] Scientific objectivity was not always the determining factor for collectors' attention. In the *Wunderkammern,* which were the forerunners of the museums of our day, everything between heaven and earth was collected and ordered more according to whim than any system. Stuffed birds were squeezed in between Chinese boxes, pickled lizards and fish tumbled in jars of cognac, inlaid jacaranda chests housed cosmetic preparations from Java and the Moluccas. The walls were hung with lances, axes decorated with feathers, shields, and porcelain.

In Sweden, the situation was somewhat different. Sweden had indeed once been a Great Power with territory around the Baltic and in northern Germany. But those days were gone. Sweden no longer harbored any military aspirations beyond her sovereign territory, and was therefore not among the nations vying for economic and military influence—Holland, England, France, and the nations of the Iberian Peninsula. Swedish involvement, as in East

Indian trade, was modest in proportion and posed no threat to the other parties. Thus Sweden was able to play a more purely scientific role. It was also more than mere coincidence that when it came to deepening understanding about distant lands in Sweden, it was the Academy and its first president, Linnaeus, who made the most substantial contribution, and not military or commercial interests.

Sweden lacked both the means and the motivation for sending out its own expeditions. Instead, up to the time of the Swedish polar expeditions in the latter half of the nineteenth century, most of the work was done by what might be called the "one-man-expedition" method. Academy support was generally granted to individuals; the infrastructure of these expeditions, in the form of vessels, equipment, and personnel, was often provided by other nations. One significant aspect of Sweden's standing as a small nation, or at best a middle-sized power, was that Swedish scholars were welcomed wherever they appeared. Linnaeus' own disciples, who were the first to benefit from Academy subventions, were greeted enthusiastically as members of the scientific avantgarde throughout the world. When 22-year-old Pehr Löfling arrived in Madrid, he was hailed as something of a grand duke of botany, which indeed he was compared to his Spanish colleagues, raised as they were on the writings of Vaillant and Tournefort.[19] Daniel Solander was a great success in London, where he assiduously tended the growing collections of the British Museum. He also made a big splash in the salons, where he was a regular guest known for his generous nature—typically, he never had time to publish a major work of his own.[20] During his first sojourn in the Cape Province in 1772, Anders Sparrman, then aged 24, was visited by Johann Reinhold and George Forster, who had just arrived with Cook's second expedition. They immediately insisted that the young Swede join them, and after a single evening's reflection he agreed, bearing in mind that for the rest of his life Linnaeus himself had regretted declining a Dutch invitation to travel to South Africa.[21] Another botanist, Carl Peter Thunberg, was on a Dutch payroll and was one of the few Europeans allowed to enter Japan—otherwise only Jesuit missionaries passed through that needle's eye. Other Linnean disciples took part in voyages under Russian (Johan Peter Falck) and Danish (Peter Forsskål) flags. These examples demonstrate the truly cosmopolitan nature of the work carried out on behalf of the Academy of Sciences.

Aside from their diplomatic neutrality another factor made the Swedes desirable partners. As students of Linnaeus, they were bearers of new, unique, and valuable knowledge. Of course, controversy still raged in the mid-eighteenth century about classification; Buffon and Tournefort were the main competitors to Linnaeus' natural classification system based on reproduction. But throughout most parts of the Scientific world, his principles were winning acceptance. His scholarship was already legendary in his own lifetime, and his position in botany (not least promoted by Linnaeus himself in *Bibliotheca botanica* (1736), was akin to that of an emperor. Being blessed with one of his disciples as an assistant was a privilege.

Travels to China

The first half of the eighteenth century, in Sweden and the rest of Europe, was notable for its rampant Sinophilia. China became known through missionary reports and trade; admiration for its teachings and culture, its legal system and institutions abounded. Interest grew in Chinese manners and customs, agriculture and economics, and also in its natural attractions. The East Indian Companies of England and Holland proved important links between a curious scientific world and the beckoning land in the Far East. Amsterdam became a center for this traffic and its botanical gardens were unparalleled in the Europe of the late seventeenth century.

An East India Company was established in Sweden in 1731, and became particularly vital and profitable by mid-century. Warm relations developed quickly between the Company and the Academy of Sciences, soon growing as intense as those of their Dutch and English counterparts. The very first year of the establishment of the Academy, 1739, saw the election of one of the Company's cargo officers, Hans Teurloen, to the Academy, at the recommendation of Linnaeus. The condition was, however, that he donate two books with Chinese paintings to the Academy. The following year Teurloen was joined by Admiral Theodor Ankarcrona, one of the Company's directors, who even contributed to the Academy's Proceedings an opus on the "five-fingered fish" of China. It seems apparent that even at that early stage, only months after the founding of the Academy, Linnaeus appreciated the value of these East India contacts.[22]

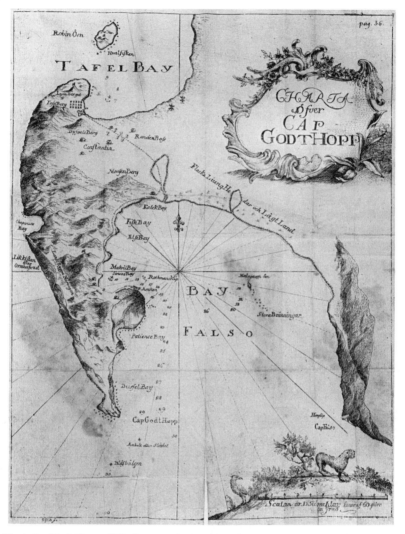

Captain Carl Gustaf Ekeberg (1716–1748), drew his own map of the Cape of Good Hope.

When the second charter was granted in 1746, research cooperation really gained momentum. It was again initiated by Linnaeus, who must have sensed from the beginning how these possibilities for geographical expansion would help reinforce his sexual classification system. For a while his ambition was to secure the right to send out travelers to China annually, paid for by the East India Company. Even though the influential Carl Gustaf Tessin handled the negotiations, no agreement was actually reached, although a vague promise was made that the wishes of the Academy would be granted whenever the opportunity presented itself. To get things under way, the Academy itself appointed a pair of scientists and secured them passage.

First out was Linnaeus' pupil Christopher Tärnström. His grim fate, to meet death in the service of scientific world conquest, would be shared by many more of Linnaeus' travelling apostles. When statistics were finally compiled in the 1780s, they showed that of the 20 travellers, eight lost their lives (most of them were no more than 30 years of age), while another went mad.[23]

Next in line was Johan Fredrik Dalman, a young mathematician who returned in fine fettle from Canton in the summer of 1749 after having assiduously collected plants, birds, fruits, and minerals, all the while dutifully keeping up the obligatory (though in his case hardly sensational) diary, dominated by meteorological observations.[24]

After 1750 cooperation took on new and more flexible forms. The driving force behind the new phase was Magnus Lagerström, littérateur and former civil servant in Pomerania before coming to work for the East India Company, first as secretary and factotum and later rising through the ranks. Science was closest to Lagerström's heart, and his essential role was coordinating the activities of the Company with the interests of the Academy— always to the advantage of the latter. Qualified professionals accompanied every voyage: priests, doctors, cargo officers. Some of these men were already interested in research, and dormant spirits could be roused by Lagerström's enthusiasm. By 1750 he had succeeded in eliciting a formal directive from the Company's executive stating that in the future, all ships' surgeons and priests must be prepared to perform scientific duties, that captains and mates must keep cartographic, magnetic, and astronomic observations in their logs, and that cargo officers must take notes on their trade dealings with the Chinese. Lagerström himself received the diaries and specimens as they arrived at the port of Gothenburg and sent them along to the Academy and to Linnaeus, who requisitioned whatever was intended for the Academy for his own private use. Since moving from Stockholm to Uppsala in 1741, Linnaeus was responsible for analyzing and describing the specimens received, though he seems to have returned to the Academy only those that were too expensive for him to take care of by himself. One large donation from Lagerström, apparently intended for the Academy, ended up in Linnaeus' possession amid loud protests from the Academy.[25] Lagerström's selfless solicitude on behalf of science was destined to be rewarded in the end. In 1748 he was inducted into the Academy, and in 1754 Linnaeus, his cup running over with gratitude, published his highly laudatory dissertation *Chinensia Lagerströmiana*.

Thus, the specimen cupboard at the Academy of Sciences was constantly being refilled with new harvests—bird nests, turtle eggs, seashells, corals, engravings and models of silk looms and bellows, maps, and beautifully illustrated volumes from Chinese printing presses. In addition, diaries and manuscripts written by the shipboard researchers arrived home with the sailors. Several of these assume a prominent place in the history of Swedish science, though they are hardly remarkable in an international context. With perhaps one exception: Pehr Osbeck's account of his travels in China awoke well-deserved admiration in many circles and was translated into German and English. It was considered important enough for Banks and Solander to include it among their meticulously selected portable library for their trip around the world on board the "Endeavour."[26]

Osbeck too was rewarded with membership in the Academy, a favor bestowed on a number of Lagerström's specimen collectors. One of these was Gustaf Fredrik Hjortberg, who kept diaries and wrote animal descriptions and who returned home with a half-tame ape from Madagascar, thus enlivening his duties as a provincial cleric. Another was William Chambers, who was born in Gothenburg of British parents, but gained fame as an English architect and subsequently was elected as a foreign member after having supplied the Academy with his writings on China, especially the great *Design of Chinese Buildings* (1757). Christopher Henric Braad spent a lot of time in China and India, but failed to assemble a work on folk traditions and trade of the sort he had long planned. As a consolation he had several essays printed in the Proceedings: one on the indigo tree (1759), another on coffee plantations in Yemen (1761), and another on the 'saga tree' (1775).

Another name deserves mention, that of Carl Gustaf Ekeberg, a sea captain who made innumerable trips to China and was inducted into the Academy in 1761 in appreciation of his service to science. His research dealt with magnetic inclination at sea, among other things, but he also brought back specimens, most notably a live tea shrub, which soon withered and died in its new environment, but which had driven Linnaeus wild with delight on arrival. Ekeberg was also his era's typical utilitarian, supporting the importation of exotic fruits and seeds for domestic production. He brought home a recipe for soya sauce from China, delivered pictures of an oil-press, and advocated the Chinese method of hatching duck eggs with the help of a warm oven.

Thus the Academy, though resting on shaky financial foundations, managed to achieve vigorous activity, at least on the surface. Of course, quantitatively the results cannot compare with the collections that were being assembled in Europe's business metropolises, but considered as the fruit of the labor of a handful of dedicated men from a little northern province of Europe, where botanical gardens were intolerant of exotica on climatic grounds, the result was impressive indeed. However, the harvest of hard scientific data was disappointing. Few conclusions could be drawn from the material crowding the shelves. The Proceedings, the foremost avenue of publication for the discoveries, boasts only a few scattered articles above those already named. The published material was fragmentary and incomplete, much like the collections, and larger, cohesive works in Swedish are not to be found.

This estimation becomes less severe if we focus our attention on botany. Linnaeus sat in Uppsala selecting all the best portions of the incoming goods, and in his case the components were brought together in a magnificent synthesis. But then again, Linnaeus' raw material was not merely limited to deliveries from the East India Company; his reach extended beyond the Far East. Linnaeus understood that as far as research on China was concerned, it was best to integrate his interests with those of the Academy. And he intended to continue along the same route.

The Travelling Apostles

With only slight exaggeration, it could be said that scientific travel in Sweden had but one originator: Linnaeus.[27] He had taken the initiative for creating the Academy of Sciences, and as soon as the opportunity presented itself, he began making a case in the new forum for scientific odysseys. It was Linnaeus who worked up the contacts with the East India Company, and he was the sole recipient of a considerable portion of all the objects, specimens, and manuscripts resulting from these voyages. Naturally, this description is not entirely just. Protocols and correspondences testify that others also went out of their way to help the travellers—Tessin's involvement is well known, but permanent secretary Pehr Wargentin also stepped in with aid when needed.

However, these deviations from the norm are marginal, the overall pattern remains the same. For a quarter of a century, from 1746 when Tärnström set out, until the 1770s, the disciples of Linnaeus not only marched in the vanguard of scientific travel, they also swelled the ranks of the footsoldiers. They were 20 souls in all, most of them young men, recruited by Linnaeus not only because he felt they might be able to pull off a scientific coup, but also with a mind to their chances of enduring the physical hardships, which would prove difficult enough—nearly half of them died in service (see note 23). Linnaeus' own enthusiasm was dampened with the years, perhaps because despite everything, the results never really justified the effort—and the price in human life. Still it was he who acted as *primus motor,* and down to the end, he eagerly awaited new results, flattering, enticing, cajoling his disciples. When Thunberg expressed doubts about his Japanese trip, Linnaeus quickly moved in with advice meant to steel the spirit and calm the nerves: "You now have a chance to make yourself renowned and immortal. . . . I am absolutely convinced that your voyages there and back will be happy ones, for I see it in a certain omen."[28] Sometimes Linnaeus operated through the Academy; at other times he sought money and contacts on his own. When he himself did not exert pressure, others at the Academy did.

What was it that made traveling as a project for collecting data so attractive to Linnaeus and the Academy during the Academy's early years? The general background has been discussed above, but a few specifically Swedish circumstances must be added, and they are all more or less bound up with the person of Linnaeus.

Linnaeus had a great personal investment in broadening the empirical base for his sexual classification system. In the 1740s, his system still had serious competition from several quarters, and he was not a man to consider the battle over and victory at hand in such circumstances. His years in Holland had laid the groundwork for world conquest, not the least through his clever marketing of both himself (preferably clad in a Lapp costume for formal occasions) and his scientific ideas.[29] Following his return to Sweden, he was ready to raise his ambitions to a global scale. Linnaeus was convinced that the study of flora and fauna on distant continents would prove the universal application and advantages of his classification model. At the same time that the disciples began their worldwide campaign, Linnaeus began work on a parallel project at home in Uppsala, the gigantic *Species plantarum* (1753), his goal being nothing less than to describe every species of plant life known to man.

This attempt at embracing the whole world matches another feature of Linnaeus' profile as a scientist: divine inspiration. Linnaeus was convinced that his discovery of the reproduction and organization of plants had revealed the blueprint God had used for His Creation. In fact, there is evidence that Linnaeus believed himself chosen by God for the task. There was surely no reason to think that God's diagram of Creation should not be valid on other continents, and thus it became a holy duty to explore this in a physico-theological spirit. Linnaeus was a son of God, and his most pressing task was to surround himself with disciples who would spread his gospel among the nations. Therefore it is appropriate to refer to these individuals as "apostles," and in fact the tradition of doing so began with Linnaeus himself in his autobiography.[30]

For Linnaeus, travel was not an abstraction; he had tried it himself. In 1732, the 24-year-old Linnaeus had set off on a summer expedition to Lapland lasting several months. It proved to be one of the most formative experiences in his life. The trip to unexplored, sparsely populated, exotic Lapland was his first and most significant. He travelled to other Swedish provinces as well in the two following decades, partly with pragmatic motives—mercantile interests needed to know if there were any valuable natural resources out there ready to be put to work for the good of the nation. But Linnaeus and his disciples were also always on the look-out for discoveries that could benefit science.[31] On these trips he came to the realization (otherwise a trivial fact for a botanist) that new knowledge could be acquired by movement in space. New species and variants could not be studied under a magnifying glass or pasted into a herbarium until the researcher himself hunted them down. They must be studied in their natural environments—their habitats must be sought out. And it was only now that travel could provide exact, fruitful results thanks to Linnaeus' own discovery of the simple, effective, and universal sexual system. His disciples were equipped with what their predecessors had lacked—a scientific method and an adequate nomenclature for arranging and describing their findings.[32] As a result, they also contributed to the spread of Linnaeus' working method. After the success of Solander and Banks with Cook, it became a tradition in the British Royal Navy to have a scientist join every expedition, as happened when Darwin and Huxley made those important journeys in the next century that led to the establishment of the theory of evolution.[33]

Movement in space, observed by Linnaeus himself and passed on to his students during constant excursions out into the surroundings of Uppsala University, became for him what might well be called an ideology, expressed in a variety of contexts but seldom with the same rhetorical consistency as in his acceptance speech upon assuming the chair of medicine in 1741, entitled "On the Necessity of Research Expeditions in Our Native Land." The speech is patriotic, and the new professor omits hardly any advantage boasted by Sweden. The speech also manifests the significance he attaches to the provincial expeditions he had undertaken; all the hardships, he states in a grandiloquent passage, are "outweighed to the fullest degree by the invaluable fruits I collected during my wanderings . . . especially . . . the gaining of more and more accumulated experience . . . to the advantage of both myself and others as well as . . . for our native land and all mankind." He warns of the danger involved in allowing the youth of Sweden to leave for Europe's admittedly first-rate institutions of learning before they had gained a solid footing on their native soil. Many a traveler returns just as ignorant as when he left, except that now he has taught himself to "babble elegant phrases and complicated words in foreign languages, or give detailed accounts and reason verbosely on theatres and play-acting and on the appropriate manner of attire in Italy, Spain, Germany and above all France." However, the most interesting aspect of his speech from our perspective is his reckoning that travel is an invaluable way of acquiring knowledge, one that can complement academies, museums, botanical gardens, libraries, and instruments. "Dioscorides himself says that he, in order to add to his experience, undertook many travels, and here and there in their writings the other fathers of medicine refer to and describe a considerable amount of their journeys."³⁴

Linnaeus had already sung the praises of travel as a form of knowledge in *Critica botanica* (1737), a work written and published during his years in Holland, when he was 30 years old. All the great botanical authorities had raved the hardships of travel. Travel is the very foundation of a successful career as a scientific scholar, and the best botanists have always traveled—even if it cost them their lives.

> *Oldenland* makes known to his own country the rare plants of the Cape of Good Hope, where he dies from the change of climate.
> *Plumier* thrice visits America, thrice he exposes himself to treacherous Neptune on the voyage there and back, each time he makes light of the severest trials of the sea and of the changes of climate, etc., if only he can collect plants: and not content with these voyages, he makes a fourth attempt and perishes.
> *Marcgravius,* not content with his American spoils, endeavours also to exhaust Africa, where he perishes.
> *Lippi* (Augustus) prefers a tour in Egypt and Ethiopia to a Doctor's degree, is tragically stabbed by assassins in Abyssinia, and perishes.
> *Bannister,* surveying the plants of Virginia and scaling a precipice, loses his foothold in a slippery place and dies from the shock of an unlucky fall.

Linnaeus spoke in the same spirit of the aged Turnefort, who traveled widely in Europe and the Orient, and Sebastien Vaillant, who died following his field studies of mosses.³⁵

One might wonder what the ideologizing of travel represents, above and beyond Linnaeus' own experiences and interests. For instance, why did he not lay more emphasis on experimental science, which after all had shown its enormous possibilities for fruitful results and which reached new heights in the scientific academies of Europe? Perhaps the answer lies in the fact that experimentation was indeed so successful in Europe. Linnaeus' speech, quoted above, contains passages that clearly show that Sweden's success as a scientific nation must be built on the distinctive character of the country, evidently its natural advantages, but also a specific way of pursuing scientific study that was different from the European way. Swedes can never hope to measure up to the advantages of Europe: "Where else . . . are there splendid and more numerous *hospitals* than in London? Where else are more beautiful *surgical operations* performed than in Paris? Where are neater *anatomical preparations* exhibited than in Leyden? Where are there more *botanical collections* than in Oxford?"[36] Linnaeus' patriotic emphasis on the Swedish way (which by the way culminated in the 1740s) was further based on a fundamental scepticism towards Europe, where his ideas were not catching on as quickly as he thought warranted.[37] So for that reason as well, he wanted to see Swedish botanical science distinguish itself as the finest anywhere.

Thus, what Linnaeus more or less consciously tried to do was establish a national style, carve out a niche for Swedish science guaranteeing its success in the world arena, and—by no means least important—win acceptance for the Linnean synthesis of the life sciences. That style was scientific travel.

Proof of this can also be found in the form these expeditions took. The scientific travelers under the Linnean banner did not set out like the heroes of which the next century boasts so many. The names of Peter Forsskål (dead at age 35 of malaria in the Yemenite mountain village of Jerim), Fredrik Hasselquist (dead at age 27 of phthisis near Smyrna), and Pehr Löfling (also dead at age 27 in the Venezuelan jungles of a fever), have not been heard round the globe, like the names of Adolf Erik Nordenskiöld and Sven Hedin. Even the more fortunate among Linnaeus' apostles, such as Carl Peter Thunberg, whose Japanese and Javanese experiences won him wide respect among the scientific community of the day (even though Guiseppe Acerbi, the Italian explorer of Scandinavia, dismissed them[38]), have not become well-known personalities, even in Sweden.

We should thus not look at the results achieved by the Linnean disciples for an explanation of their anonymity, but to another phenomenon altogether. It was never intended that they draw attention to themselves; their journeys were never meant to be understood in the context of individual fame. The scheme of which they were a part ought instead to be seen as determined by two main concepts: the plan of Creation and bureaucratic loyalty. In certain instances we can add a third concept, one quite familiar to the Academy: economic benefit.

The plan of Creation was the overriding concept. For Linnaeus and many of his contemporaries, seeking out the order with which God had infused Creation lay at the center of scientific investigation. This order lacked mystical features—it was completely rational and in principle comprehensible. Linnaeus thought he was on the right track with his classification system. His research could be described as a continual laying bare of greater and greater parts of Creation, and the more that was uncovered, the more wonderous it all appeared. Thus, with constant praise of the Lord on their lips, Linnaeus and his disciples set about their work.

Anonymity and uniformity were distinctive features of the research scheme. Of course, during the eighteenth century the coming scientific ideal of expressing oneself according to a strict nomenclature and without a lot of individual fuss began to gain currency. But that alone does not provide a satisfactory explanation. Among the great majority of the disciples, a style dominated that reflected the nature of their assignment: the collection of facts. Linnaeus himself struck the keynote during his provincial tours, which are written up in an economical, technical, and unsentimental prose. This is not to imply, however, that his writing lacked rhetorical effectiveness. It is barren but beautiful. In physico-theological fact-finding, every scrap of information is of the utmost importance, not as a sign of the individual researcher's diligence and merit, but as proof of divine order.

This then is the background one should have in mind when reading the journals and travel accounts of the Linnean apostles. They are chock-full of facts and details, but as a rule lack any epic form. They enumerate and tabulate; no place or person is left unnamed. Hardly a day passes that is not considered worthy of mention, even if it offered no more than a dreary transport between one place and the next.[39]

When the third volume of Pehr Kalm's immense work on his journey to America was published, the academic world groaned under the weight of what seemed no more than verbose repetitiveness. Even the sympathetic Wargentin, who struggled mightily to see the work through financially, thought Kalm had gone too far. One circumstance that deserves mention at this point is that some of the journals could not be edited properly, since the author in question had died as the result of his expedition. Both Löfling and Hasselquist would probably have produced richer writings

than the piecemeal letters and diaries edited and published by
Linnaeus, who was certain that they were nonetheless of great
value.[40] Had Johan Peter Falck not been beset by his demons and
had retained his sanity after his journey to Siberia with Orlov (not
sponsored by the Academy), he was likely to have produced a
much finer crop of writings than the few unremarkable booklets
published by the German apothecary Johan Gottlieb Georgi in the
latter's mother tongue as "Beyträge zur topographischen Kennt-
nis des russischen Reichs" (St. Petersburg, 1785–1786).

But we must not exaggerate the idea of the disciples as inheri-
tors of their master's ideal of reporting. These travellers had ventured into an anthropological free-for-all, where they were
fascinated by curiosities and formulated racist biases, suffering
from homesickness and travel fever in a very human, perhaps
better still, very European way. Pehr Osbeck's observations about
the Chinese of Canton are no different from thousands of others
in the same genre. He needed only a few months to acquire the
usual prejudices without the nuisance of bothering to formulate
any basis for his conclusions: "They are friendly in company, sober
and neat in habit, industrious in the community, given to handicrafts and especially trade; but in addition loud, greedy, corrupt,
obstinate, conceited and distrustful."[41] Weaknesses were discovered in their writings even in their own day. The *Gentleman's
Magazine* of London remarked acidly on Hasselquist's report
(based on the age-old nemesis of truth—hearsay) that Egyptian
women hatched chicken eggs by storing them for several days in
their armpits: Was it really "worth their while to be idle so long,
merely to hatch eggs?"[42] Pehr Kalm delivered with a completely
straight face the false story of how the North American bear killed
cattle: "He bites a hole into the hide, and blows with all his power
into it, till the animal swells excessively and dies; for the air expands greatly between the flesh and the hide."[43]

Anders Sparrman distinguished himself from the general pattern. His fate was remarkable and hardly Linnean. Swallowed up
by the undercurrents of the Enlightenment, Sparrman became a
dedicated follower of Swedenborg, and made a strange trip to
Senegal, where, in the company of several like-minded friends, he
attempted to found a "Republique of God" in the spirit of his
master.[44] Inspired by his African experiences, he immersed himself in a passionate absolutism and, assuming the mantle of social
consciousness, became a doctor to the poor in Stockholm, attempt-

View of Canton with the flags of different countries flying over their warehouses. Drawing by Carl Gustaf Ekeberg, sea captain and, also, a capable amateur scientist.

ing to cure his patients with the help of Franz Anton Mesmer's "animal magnetism," which he practiced in his own particular variant.[45]

Sparrman was enormously gifted, but his career in science was a rocky one. He was mostly active on his own, and his contacts with the Academy did not become intense until after his return from his visits to the Cape and his spell as a member of Cook's second around-the-world voyage in 1772–1776. Then, for a time, he became much more diligent, writing articles in the Proceedings and delivering specimens and objects to the Academy's collection (among them a foal from the now extinct South African *quagga*), whose curator he became in 1780.[46] Thereafter, things did not proceed as smoothly; he was relieved of the collection in 1799 and lost his chair in natural history; later he resigned his professorship in surgery at the Karolinska Institute in Stockholm. "Boredom is being a bureaucrat," he sighed in a letter, thereby denouncing one of the cornerstones on which Swedish scientific expeditionary work rested.[47]

Although Linnaeus had sponsored his trips, Sparrman was not manipulated by his former teacher.[48] He was more independent than his predecessors, and perhaps that is why his diaries are written in a much more lively manner, more personal, filled with irony and humor, though he can also be severe, as when he harshly censures a poor Swedish translation of the English description of the expedition. Sparrman had an eye for spectacular and peculiar phenomena among "the natives"; however, we cannot claim that he "understood" them in any modern anthropological sense. It is also worth noting that he wrote at a much later date than the others, when a more romantic ideal held sway. Sparrman also had trouble financing his magnum opus, which, with his typical lack of haste, he would take nearly half a century to complete—the trip began in 1772, and the final volume came out in 1818.

Otherwise the apostles worked under the close supervision of their master. They received specific instructions from him, and the prospects for personal creativity were limited. Linnaeus had a habit of making up lists of what had to be done on a particular journey, and these were comprehensive. The directives Löfling received before his trip to Spain, dated May, 8 1751, covered 27 points and embraced everything from quadrupeds, birds, amphibians, fish, insects, and worms ("everything very thoroughly"), to trees, herbs, grasses, and mosses, including soil types, cultivation, uses as utility plants, and the local names of the various plants. On top of all this, Löfling was expected to find out facts concerning economics, geology, disease, and house medicines, enlist some Spanish students as assistants, deliver a "herbarium with all the herbs of Spain," and "remit a complete *Flora* and *Fauna Hispaniae.*" Finally, as if to dispell any doubt about what it was that stood on the program, Löfling was reminded to "ask the reason for everything." In economics, Linnaeus pointed out that his own "previously published journeys could further represent what may especially serve economy."[49] The instructions to Tärnström (who was to collect nutmeg and goldfish, plus make thermometric observations "night and day") and to Kalm were similarly exhaustive.[50] In instances where the instructions have not been preserved, we can assume that Linnaeus delivered them personally, or that the document has been lost. When the instructions were not followed to the letter, Linnaeus could become irritated. When Kalm, far away in America, became too independent (maybe even a bit uncertain before more protracted excursions), it amounted to insubordination in the eyes of the master. So centralized and to a certain degree methodical was this gigantic scientific maneuver.

At the same time, this provided the scientific expeditions with a modern character. It was a question of teamwork, the overall result being what counted. And it was Linnaeus and the Academy who were ultimately responsible for seeing to that. The travelling disciples were in that respect scientific bureaucrats, answering the call of the Cause, the King, and (ultimately) God. They were expected to live by the bureaucratic code of honor, displaying self-sacrificing zeal in carrying out their duties—and as a rule, they did. They were typical utilitarians. Keeping an eye open for anything that could economically benefit the Fatherland was second nature to them—Linnaeus wrote it into his instructions; the Academy used it as a selling point whenever it went begging for funds. This was especially true for Kalm's trip to America. At the behest of the Academy, Linnaeus drew up a prospectus in which he listed all the invaluable plants Kalm might be able to import from the New World: oaks, walnuts, grass seed, species of pharmacological interest, mulberry trees for sericulture. That these specimens would flourish in the Swedish climate was beyond doubt; *Radix Ninsi,* a Canadian medicinal herb, "could thrive as easily as the common columbine in our gardens."[51] Kalm himself was just as eager to stress these advantages, and fought tirelessly to fulfill all that was expected of him, sending hundreds of different root and seed species home to Sweden. There, gardeners in Lund and Uppsala, Linköping and Västervik attempted to acclimatize them to Nordic conditions, but it did not work out well. No economic successes were achieved at any rate. On his own, Kalm, who had assumed a chair in economics in Turku, succeeded in starting a plantation for walnut and mulberry trees and American plums. However, nothing came of these efforts either in the long run. Only a few ornamental plants, such as the Virginia creeper, remain as lasting reminders of these efforts at transplantation.

Destinations

What sort of travels did these bureaucrat-apostles undertake? Where did they go?

Linnaeus' disciples visited all the known continents. Kalm, whose trip was probably the Academy's most extensive project, got quite a lot done in North America despite misgivings. He visited Delaware, New York, and Philadelphia, where he met Benjamin Franklin. Two excursions to Canada took him to Montréal, Québec, and Ontario, where he saw Niagara Falls. After coming home he published *En resa till Norra Amerika* (A Trip to North America) in three volumes (1753–1761).

Fredrik Hasselquist headed for the Near East: the Levant, Egypt, and Palestine. The project was private, organized by Linnaeus and Hasselquist together, and partly financed with borrowed money. True Linnean that he was, Hasselquist presented the Academy with one discovery after another. These were often considered valuable enough to be published in the Proceedings, and one essay, on salammoniac, was published in *Philosophical Transactions* at the behest of Linnaeus in 1760, which shows that the disciples could contribute to the international reputation of the Linnean school. The Academy rewarded Hasselquist's efforts by making him a member on New Year's Day, 1751. And when his economic worries increased the next year, Academy members passed the hat among themselves. Hasselquist died in February 1752 in Smyrna, of persistent consumption. He died deeply in debt, but left behind an admirable inheritance of scientific riches—herbaria, a mummy, nature specimens, manuscripts—

which would be lost if the Academy failed to cover his debts, some 3,000 piasters. Anders Rydelius, Swedish consul in Smyrna, was put in charge of the transaction (ironic in that he was the largest claimant). The amount could not be raised until wealthy merchants and Queen Lovisa Ulrika herself had been won over to the cause, but in 1754 the treasures arrived at the Royal Palace in Drottningholm. From there, they made their way to the Royal Swedish Library and the Museum of Natural History, both in Stockholm. Linnaeus retained the letters and journals, and in 1757 published a gigantic volume entitled *Iter Palaestinum.*

As early as 1750, at the tender age of 21, disciple Löfling had been prepared to set out on a journey to the East Indies arranged by the Academy. When he finally embarked one year later, he ended up in Spain instead, without any direct instructions from the Academy. But he was after all an associate, and he provided the Proceedings with reports on corals. During his lengthy stay in Madrid, waiting to embark for his actual destination, South America, Löfling wrote long and comprehensive letters to Linnaeus, his parents, relatives, colleagues, and fellow students (in toto approximately 150 letters). Among his correspondents was Daniel Solander. Löfling spoke of botanical excursions, of saffron and mulberry cultivation, and he complained unendingly of the drought that was causing plant life to wither and himself to become depressed. He was in better spirits when he was able to work and had discoveries to write about. He also traveled in the provinces, though not as extensively as he wished. He followed the Linnean program faithfully and willingly. These excursions provided him with empirical confirmation of things that he had only heard about before. On "Coccinell" and other coloring agents of the plant world, he writes that the peasants dye clothing yellow using " 'umbella of Thapsia'. . . . There is much that I know, but do not rely on any of it, for I have hardly travelled at all."[52]

Löfling is an excellent example of how the Swedish scientist moved smoothly through the most diverse diplomatic circumstances. From a Spanish standpoint, his most important task was joining José de Iturriaga's border-marking expedition to South America to collect information about its natural assets. Löfling was intent on finding new medicinal plants and writing the natural history of the continent. Neither his Lutheranism nor the fact that other Swedes helped the Dutch reach South America caused him any problems.[53] The evidence indicates that Löfling was prepared

Carl Peter Thunberg (1743–1828). He published his *Flora Japonica* in 1784.

Anders Sparrman (1748–1820), well known for his South Sea voyages with Captain Cook.

to do his best while in the service of Spain, and he was busy with scientific work up to the end, when fever finally took its toll.

Apparently Linnaeus worried that he would not be able to partake of everything Löfling managed to collect. His farewell letter to his beloved disciple (when the latter was preparing to leave Madrid) betrays as much: "My dear Loefling, think of me when you come to your country. Send me a twig of some rare herb from your Paradise, that I might share in your delight. . . . If you are thrifty, you will surely come home with 1 barrel of gold, but I prize the specimens more. My dear, favor me always when you can with something to amuse Europe with. . . ."[54] What perhaps strains the credibility of Linnaeus' elegy (printed in the foreword to the gleanings he published as *Iter Hispanicum*, 1758), is the ritualistic character that reappears in his letters to all the disciples. Each and every one of them is partaking in the most important scientific event of the day; each and every one of them is in his own "Paradise." Linnaeus coaxes, flatters, pleads, seduces—his personal investment in the results is impossible to miss.

What Löfling had not succeeded in achieving, Linnaeus hoped that the much less promising Daniel Rolander might complete. In 1755 Rolander was sent to Paramaribo in South America, but his is a short and sad story. Rolander failed to master the physical and emotional strain. He returned home after only six months, and his burgeoning mental infirmity had suffered the worse from these experiences. The Academy, which had received a contribution on Doliocarpus in the Proceedings from Rolander, awarded him a 360 *daler* subsidy upon arrival, but there was nothing they could do to forestall madness.

Expeditions to the North

Their journeys did not always take the disciples to the ends of the earth. As early as 1739, Mårten Triewald, one of the Academy's founders, suggested that the Academy send researchers to ascertain the natural histories of the Swedish provinces.[55] Even though little happened at first (Linnaeus' provincial travels were not followed up until decades afterwards), Triewald would eventually see his suggestion fulfilled. In 1780 Johan Daniel Lundmark, another pupil of Linnaeus, travelled to Lapland with the Academy's support to collect plant and animal life.[56] He was accompanied by

Clas Fredrik Hornstedt, a zoologist (who shortly thereafter sailed to China as the last of the Academy's long-distance travelers), and the very young Olof Swartz, later famous for his studies of West Indian flora, culminating in the three-volume *Flora Indiae occidentalis* (1797–1806).

It was no coincidence that it was a Lapland excursion that received support. The regions of the far north were subject to a scientific exoticism that in certain respects is reminiscent of the scientific curiosity about distant continents. In 1736–1737, a French survey crew headed by Maupertuis and intent on determining the shape of the earth once and for all had travelled to northern Scandinavia. The Swedish geodesist Jöns Svanberg repeated their experiment around 1800, reporting his findings to the Academy.[57] Also worth noting are the Lapland travels of natural historian Göran Wahlenberg, the first of which he undertook in 1802 with partial funding from the Academy. He covered large expanses of the Arctic region and wrote a dissertation on the environs of Kemi in Finland.[58] The second trip, now fully backed by the Academy, took place in 1806. Wahlenberg studied flora and glaciers and made height and temperature observations of the mountain regions in latitude 67° north.[59] Lars Levi Laestadius, botanist and renowned preacher, also enjoyed Academy support when he studied the plant geography of the most northerly regions of Sweden in 1824.[60]

Lapland enjoyed special treatment, but most of the provinces were eventually visited by inquisitive scholars. Several more trips won the support of the Academy, including two to the island of Öland.[61] Some travel accounts were published in the Proceedings, even if undertaken without Academy involvement.[62] In the early nineteenth century, we witness an increase in interest in the provinces, partly resulting from the Romantic era's interest in regional distinctiveness, partly because of continued feverish activity in botany, and partly (as we reach the middle of the century), through the survey and charting of Swedish geology.[63]

Thus, scientific travel continued unabated, though in new forms. Globe trotting in the spirit of Linnaeus was a closed chapter, and a reorientation toward the Swedish and the provincial had begun. Yet the northern orientation of this nineteenth-century research had one dimension that assured Sweden a place on the international stage—polar research. Polar research had also been initiated by a disciple of Linnaeus, Anton Rolandsson Martin, who spent three trying months drifting aimlessly about in the northern Arctic Ocean. Unfortunately, he made only short forays onto Spitzbergen, and came home nearly empty-handed.[64] Still, it was a beginning, and when zoologist Sven Lovén resumed modern Swedish polar research in 1837 with the support of the Academy, it was both a new beginning and the continuance of a tradition (cf. Tore Frängsmyr's article in this volume).

The geographic orientation had been reversed—away from the Tropics and the Southern Hemisphere and toward the Arctic north. But as always science remained the central motive for these journeys. Commercial, political, and strategic motives had informed the exploration of distant continents by the European powers during the Renaissance and the Enlightenment, and the same motives lived on in the European attempts at approaching the polar regions. The little northern realm of Sweden was forced to rely on a more modest, but in the long run no less vigorous, motive—science itself.

Notes

1. Günter Schilder, "New Holland: The Dutch Discoveries," in *Terra Australis to Australia,* eds. Glyndwr Williams and Alan Frost (Melbourne, 1988), 83–115.
2. Alan Cook and Clive Holland, *The Exploration of Northern Canada, 500 to 1920: A Chronology* (Toronto, 1978), 31–48.
3. Sverker Sörlin, *Framtidslandet: Debatten om Norrland och naturresurserna under det industriella genombrottet,* (Stockholm, 1988), 23–27 and literature cited therein. Summary in English: "Land of the Future: The Debate on Norrland and Its Natural Resources at the Time of the Industrial Breakthrough."
4. The clerical reports on which Schefferus based his work are reprinted in *Berättelser om samerna i 1600–talets Sverige,* facsimile edition (Umeå, 1983).
5. Jonathan Swift, "On Poetry: A Rhapsody" (1733), quoted here from *Jonathan Swift: The Complete Poems,* ed. Pat Rogers (Harmondsworth, 1983), 526.
6. Francis Bacon, *Essays* (1625), (London, 1973 and subsequent editions), 54 ff.
7. Barbara Maria Stafford, *Voyage into Substance: Art, Science, and the Illustrated Travel Account, 1760–1840* (Cambridge, Mass. 1984), 43.

8. J. Bertrand, *L'académie des sciences et les académiciens de 1666 à 1798* (Paris, 1869), 108–145. Ch. R. Weld, *A History of the Royal Society*, vol. 1 (London, 1884), 351 f. Cf. Lindroth, *KVA Historia*, here vol. 1:2, 630. Lindroth's history of the Academy and Tore Frängsmyr's on the East India Company, *Ostindiska kompaniet: Människorna, äventyret och den ekonomiska drömmen* (Stockholm, 1976), contain much general information about the Academy and the scientific travellers of the second half of the eighteenth century. I wish to stress here that such information will appear throughout this essay, without specific reference to these works.

9. R. W. Frantz, *The English Traveller and the Movement of Ideas: 1660–1732* (Lincoln, Nebraska, 1934), 22 f. Cf. Bernard Smith, *European Vision and the South Pacific* (1960), 2nd ed. (Sydney, 1984), 8.

10. T. W. Freeman, *A Hundred Years of Geography* (London, 1961), 50.

11. Sten Lindroth, "Svensk-ryska vetenskapliga förbindelser under 1700-talet," in his book *Löjtnant Åhls äventyr* (Stockholm, 1967), recounts the repeated attempts at pressing eastwards.

12. *Navigantium atque itinerantium Bibliotheca*, ed. John Campbell, 2 vols. (London, 1744–1748); here vol. 1, 332; quoted in Alan Frost & Glyndwr Williams, "Terra Australis: Theory and Speculation," in *Terra Australis to Australia*, 28.

13. Bernard Smith, "The Intellectual and Artistic Framework of Pacific Exploration in the Eighteenth Century," in *Terra Australis: The Furthest Shore* (Sydney, 1988), 123–127.

14. Patrick O'Brian, *Joseph Banks: A Life* (1987), (London, 1988), 59 ff., 156 ff.

15. Schilder, 83.

16. Lucile H. Brockway, *Science and Colonial Expansion: The Role of the British Royal Botanical Gardens* (New York, 1979).

17. William Eisler, "Terra Australis: Art and Exploration 1500–1768," in *Terra Australis; The Furthest Shore*, 25 f.

18. Sten Lindroth, *Svensk lärdomshistoria*, 4 vols. (Stockholm, 1975–1981); vol. 3 (1978), 176.

19. See, e.g., letter from Pehr Löfling to Linnaeus 4/11 1751, Linnean Society, London, Linnaeus' letters, Vol. IX, 398 f. For further information on Löfling, see Stig Rydén, *Pedro Loefling en Venezuela (1754–1756)*, Instituto Ibero-Americano (Gotemburgo and Madrid, 1957). An English version of Linnaeus' posthumously printed collection of the papers Löfling left behind, *Iter Hispanicum* (Stockholm, 1758), was published by Jean Bernhard Bossu, "An Abstract of the Most Useful and Necessary Articles Mentioned by Peter Loefling," in *Travels through That Part of North America Formerly Called Louisiana* (London, 1771), vol. 2, 60–422.

20. Roy Anthony Rauschenberg, "Daniel Carl Solander: Naturalist on the

'Endeavour,' " *Transactions of the American Philosphical Society* 58:8 (1968), 5–66.

21. Anders Sparrman, *Resa till Goda hopps-udden, södra polkretsen och omkring jordklotet,* in two parts, part two comprising two volumes (Stockholm, 1783–1818); here vol. 1, 87–90. The first part has been published in an English translation: Andrew Sparrman, *A Voyage to the Cape of Good Hope, towards the Antarctic Polar Circle and round the World: But Chiefly into the Country of the Hottentots and Caffres, from the year 1772 to 1776* (London, 1785).

22. KVA Protokoll, 5/9, 28/11, 1739.

23. Among the dead were Carl Fredrik Adler (in Java at the age of 41); Andreas Berlin (Guinea, 27); Johan Peter Falck (Kazan, Russia, 31); Petrus Forsskål (Yemen, 31); Fredrik Hasselquist (Smyrna, 27); Pehr Löfling (Venezuela, 27); Olof Torén (after a journey to East India, 35); Christoffer Tärnström (Pulo Candor, 37). Daniel Rolander went mad during a trip to Guyana. The remaining eleven were: Adam Afzelius, Johan Fredrik Dalman, Gustaf Fredrik Hjortberg, Pehr Kalm, Mårten Kähler, Pehr Osbeck, Anton Martin, Göran Rothman, Daniel Solander, Anders Sparrman, Carl Peter Thunberg.

24. Johan Fredrik Dalman, "Dagbok under resan från Giötheborg till Kanton och hem 1749," KVA Archives.

25. Linné, *Bref och skrifvelser,* 2 parts in 8 volumes (Stockholm, 1907–1943); here I:2, 141. Cf. Yngve Löwegren, *Narualiekabinett i Sverige under 1700–talet: Ett bidrag till zoologiens historia,* Lychnos-bibliotek 13 (Uppsala, 1952), 198 ff. 266 f.

26. Pehr Osbeck, *Dagbok Öfwer en Ostindisk Resa Åren 1750, 1751, 1752* (Stockholm, 1757); Smith, 127.

27. A survey of Linnaeus' different activities is presented in *Linnaeus: The Man and His Work,* ed. Tore Frängsmyr (Berkeley, 1983).

28. Linnaeus to Carl Peter Thunberg 1/11 1771, Bergianska brevsamlingen, vol. 19, 520, KVA Archives.

29. Gunnar Eriksson, "The Botanical Success of Linnaeus: The Aspect of Organization and Publicity," in *Linnaeus: Progress and Prospects in Linnaean Research,* ed. Gunnar Broberg (Stockholm and Pittsburg, 1980), 57–66.

30. Lindroth, *Lärdomshistoria,* III, 233 ff.; Wilfrid Blunt, *The Complete Naturalist: A Life of Linnaeus* (Collins: London, 1971), 183. Cf. Frängsmyr, "Linnaeus as a Geologist," in *Linnaeus, the Man and His Work,* 143–155.

31. Linnaeus's travel accounts of his provincial journeys came out 1734 (Dalecarlia), 1745 (Öland and Gotland), 1746 (Västergötland), 1751 (Scania).

32. William T. Stearn, "Botanical Exploration to the Time of Linnaeus," *Proceedings of the Linnean Society of London* 169 (1958), 175.

33. Blunt, 184.

34. Carl von Linné, "Om nödvändigheten af forskningsresor inom fäderneslandet," Swedish translation of the Latin original, in *Skrifter af Carl von Linne*, 6 vols., (Uppsala, 1905–1913), here vol. 2 (1906), 69–72.

35. Linné, *Critica botanica* (1737), English translation *The "Critica botanica" of Linnaeus* (London, 1938), 65 ff.

36. Linné, "Forskningsresor," 71.

37. Frans A. Stafleu, *Linnaeus and the Linnaeans: The Spreading of Their Ideas in Systematic Botany, 1735–1789* (Utrecht, 1971), chapters 8 and 9.

38. Giuseppe Acerbi, *Travels through Sweden, Finland, and Lapland to the North Cape, in the Years 1798 and 1799*, 2 vols. (London, 1802), I 121.

39. Spectacular examples are offered by Osbeck, *Dagbok*, and Forsskål, *Resa till Lycklige Arabien: Dagbok 1761–1763*, ed. Arvid Hj. Uggla (Uppsala, 1950), 125 ff.; types of currency, prices of products, and units of measurement are accounted for in minute detail.

40. Fredric Hasselquist, *Iter Palaestinum eller Resa till Heliga landet*, (Stockholm, 1757); *Petri Loefling Iter Hispanicum, eller Resa til spanska länderna* (Stockholm, 1758).

41. Osbeck, *Dagbok*, 169.

42. Quoted in Percy G. Adams, *Travelers and Travel Liars 1660–1800* (Berkeley, 1962), 227 f.

43. Kalm, *Travels in North America*, ed. John Reinhold Forster (London, 1770–1771), vol. 1, 116 f.; cf. Adams, 232 f.

44. Ronny Ambjörnsson, "Guds Republique: En utopi från 1789," *Lychnos* 1975–1976, 1–57. More on contacts made in connection with the journey, impressions of Paris, etc., can be found in Sparrman's correspondence, UUB.

45. Karin Johannisson, *Magnetisörernas tid: Den animala magnetismen i Sverige*, Lychnos-Bibliotek 25 (Uppsala, 1974), 221 ff. [Anders Sparrman], *Om prestmedicin och animal magnetism* (Stockholm, 1815). For this publication, as well as for his entire magnetic practice, Sparrman was held up to ridicule in many quarters, e.g., by J. Berzelius, "with his Chemical arrogance and medical ignorance"; Sparrman to Carl Peter Thunberg, no date, UUB G 300 ä.

46. Which he also notes in his diary: Sparrman, *Resa*, II:1 (1802), 84; II:2 (1818), 61, 74. On the reception of the collections in Sweden, see KVA Protokoll 11/9 1776.

47. Anders Sparrman to Claes Alströmer 25/4 1785, UUB, G 6:61.

48. Linnaeus to the Academy, no date, transcript recorded in the minutes 13/11 1771, KVA Protokoll.

49. Linnaeus May, 8 1751, quoted in Stig Rydén, *Pehr Löfling: En lin-*

nélärjunge i Spanien och Venezuela 1751–1756, (Uppsala, 1965), 29 f.

50. Instructions to Tärnström reprinted in *Bref och skrivelser,* I:2, 53 f.
51. Linnaeus' memorial 10/1 1746, in *Brev och skrifvelser,* I:@, 58 ff.
52. Löfling to Gustaf Lidbeck July, 16 1753, quoted here from Rydén, *Pehr Löfling,* 80
53. Axel Paulin, "Skeppet 'Fortunas' expedition till 'Wilda kusten af Sïdra Amerika': Ett bidrag ur Sveriges kolonialhistoria," *Forum Navale* 1951; Rydén, *Pehr Löfling,* 148, 166 f.
54. Linné, *Bref och skrifvelser,* I:3, 310.
55. KVA Protokoll 15/12 1739.
56. *Ibid.,* 3/5 1780.
57. Jöns Svanberg, "Berättelser öfver resan til Pello," in *KVAH* 1799; *idem,* "Historisk öfversigt af problemet om Jordens figur," *ibid.,* 1804.
58. Göran Wahlenberg, *Geografisk och ekonomisk beskrifning om Kemi lappmark i Vesterbottens höfdingedöme* (Stockholm, 1804), 4.
59. Göran Wahlenberg, "Anmärkningar om Lappska vegetationen, med beskrifning om *Myosotis deflexa* eller ett nytt förgätmigej från Lappland," *KVAH* 1810, and *Berättelse om mätningar och observationer för att bestämma lappska fjällens höjd och temperartur vid 67 graders polhöjd* (Stockholm, 1808); on the Academy's support for the expedition, see *ibid.,* 5.
60. Lars Levi Laestadius, "Beskrifning öfver några sällsyntare Växter från norra delarne af Sverige jemte anmärkningar i Växtgografien," *KVAH* 1824, 160; continuation *ibid.,* 1826,
61. Cf. *ibid.,* 1813, 131; 1816, 166.
62. E.g., Carl Joh. Hartman, "Physiographiska observationer under en Resa genom vestliga delarne af Gestrikland, Helsingland och Jämtland," *KVAH* 1818, 121–160. The majority of the Swedish provinces were paid at least some attention, even though the references to Lapland are unusually numerous; A. J. Ståhl, *Register öfver Kongl. Vetenskaps-academiens handlingar ifrån deras början år 1739 till och men år 1825* (Stockholm, 1831).
63. For an overview, see Elof Colliander's bibliography *Kungl. Svenska Vetenskapsakademiens skrifter 1826–1917* (Stockholm, 1917), esp. 261 ff. (botany), 227 ff (geology).
64. Martin's diary is reprinted in *Ymer* 1 (1881), 102–141.

WILHELM ODELBERG

Berzelius as
Permanent Secretary

APART from Carl Linnaeus, no Swedish scientist has been the subject of so many portraits as Jacob Berzelius. Even the iconography itself has been portrayed—in an excellent work published by the Academy of Sciences at its bicentenary in 1939. At the same time, Berzelius was paid the honor of being depicted on a stamp, as he was again in 1979.[1]

Among the first pictures with Berzelian connections is a portrait that has acquired the name "the hungry Berzelius." It shows a young man, sitting with a book at a table, on which there is a stand with a retort. He is dressed in a director's frock coat, in other words the garment that emerged from France soon after the French Revolution, blue and white knee-breeches, white stockings, and heavy shoes. The portrait is undated and unsigned. In the profile, in the area of the cheek and the nose, it is consistent with later ones. The figure is lean and delicate—possibly indicating malnutrition. In the summer of 1802, Berzelius had been appointed royal physician at Drottningholm. Typhoid fever was raging in the surrounding area. Berzelius caught it himself and was for a while at death's door. After he recovered, his whole constitution seemed to change. The previously delicate 23-year-old rapidly became a well-built man, even a somewhat portly one. During later periods of illness in Berzelius' life, he wrote in letters that lack of appetite had led to loss of his paunch. Be that as it may, the portrait depicts a young scientist who has seen his share of trouble. It has sometimes been pointed out that until Berzelius, by the fortuitous—for him—unexpected death of the incumbent, obtained a professorship in Stockholm, the unrelenting fates seemed to be conspiring to lay an endless succession of obstacles in his path. His father died when Berzelius was small; his mother remarried, but she, too, died soon after. His stepfather, Ekmarck, then married again. The boy was made aware by his new stepmother that he was inconvenient and unwanted. From then on he was cared for by uncles and aunts on his mother's side and by other close or more distant relatives.[2] He had a particularly hard time with his mother's family, the Sjöstens, at Väversunda. It was this that Berzelius had in mind when he later wrote in his autobiographical notes, "I have never been able to take the same pleasure as others in recollections of childhood."[3]

The poor boy's schooldays in Linköping were not happy either. It was here that his compelling passion for science first started to shape his life. The great majority of the boys would become clergymen—that was a strong tradition at the school. He had to choose between the church and medicine. The former certainly provided a reliable living, often quite a comfortable one. A doctor's prospects were at this time much less certain. Medical services were primitively organized in the late eighteenth century, and beds were few. The young Berzelius specialized strongly, studying insects and birds with intense enthusiasm. He liked roaming the countryside, with or without his gun. His attendance at lessons in subjects that did not engage his interest was reluctant.

His obvious interest in science was encouraged by the strange Linnean and physician Claes Fredrik Hornstedt, who was a master at the gymnasium. Hornstedt's attitude was in marked contrast to that of the headmaster of the school, who so disapproved of Berzelius' various doings that the young man came near to being ignominiously birched. Berzelius' final report from the gymnasium was not really as poor as he himself later made out in his autobiography, whose unreliability on matters of fact is strange considering the meticulous nature of its author as a scientist. The diocesan superintendent of schools, Jacob Lindblom, raised an admonishing finger: "I know that you have been absent from many lessons, but I know also you have not employed the time badly—employ your gifts, and you will be a useful citizen."[4]

It might well have crossed Lindblom's mind that he himself would one day become an archbishop, but it could hardly have occurred to him that his problem pupil would attain international renown in an as yet relatively undeveloped science and, indeed, be ennobled. Money was, of course, a constant problem to the young Berzelius. Like any talented but impecunious youth, he had to become a tutor in more affluent families. Moreover, he was unsure of his true vocation when at last he found himself able to live in Uppsala. A doctor is what Berzelius became, but it was chemistry that gradually claimed his entire interest. It was not Berzelius' fault that he became caught in the crossfire of a quarrel between the professor of chemistry, the eccentric Johan Afzelius, and the celebrity of the medical faculty, Adolf Murray. The squabble between the two professors might easily have caused Berzelius to move to Lund, but matters were finally settled, and he got his medical degree. Nevertheless, laboratory work was Berzelius' world, both in the chemical laboratory at the university and in his own home, where he could arrange his experiments on a little stove. In 1802 he was ready to enter the big world waiting for him outside the university. From Berzelius' almost 50-year career as a chemist, his electrochemical theory has been identified, probably rightly, as the outstanding achievement. As we shall see, he accomplished much else besides. Berzelius' career as a mature man spanned the period between two world-shaking events. When he first made his appearance, the violent aftermath of the French Revolution had still not subsided. When he died, the reverberations of the February Revolution, with its storm center in Paris, were being felt over most of Europe, even in the then backwater of Sweden.

Chemist and Professor

Lavoisier brought a new order to chemistry at the end of eighteenth century. When Berzelius arrived on the scene, the situation was somewhat confused, and the concepts of the atom, equivalence, and the molecule had not properly stabilized. In France the antiphlogistic approach of Lavoisier has been called "la révolution chimique." In place of the hypothetical phlogiston, the fire element, which was assumed to leave a body during combustion, oxygen, the substance recently discovered by Carl Wilhelm Scheele and Joseph Priestley, was suddenly placed in the center of the whole chemical system. This also meant that the language of chemistry changed, together with all its terminology. Older

The "hungry" Berzelius. Water colour by an unknown artist, ca. 1810.

chemical works could not therefore be understood at all by the nineteenth-century scientist unless he had a particular knowledge of history. From the vanquished phlogiston theory, one element survived, but an important one—the dualistic principle. This stipulated that every chemical compound was composed of two simple constituents with mutually opposed properties, phenomena that we recognize from conceptual systems devised much later.

The influence of Lavoisier was soon to dominate the science of chemistry. His immediate successors did various things to make his theories more widely known, but they hardly developed them. Meanwhile, in physics, Galvani and Volta were reporting epochal discoveries. This is a particularly apt term to apply to the work of Volta, who in 1800 invented the electric battery in the form of the Voltaic pile, which has been described as a worthy christening gift to the new century. It was to play an important part in Berzelius' future work. After his disputation on the medical effects of galvanic currents, Berzelius had joined an exclusive little circle in Stockholm that called itself the Galvanic Society. Here he met the industrialist Wilhelm Hisinger, and a scientific collaboration soon began between them that was to last for the rest of Berzelius' life. Unfortunately most of Hisinger's many letters to Berzelius have been lost. The most likely explanation for their disappearance, considering that the correspondence in the opposite direction has been preserved, is that Hisinger asked for the return of his letters after Berzelius' death and burned them.[5] Berzelius has reported on the activities of the new friends in his autobiography. Hisinger had set aside several rooms in his house (at the corner of the present-day Nybrogatan and Riddargatan), a seventeenth-century building also known as the German Baker's House after its first owner, Mårten Kammecker. Berzelius says:

"Hisinger was engaged on hydroelectric experiments and we at once undertook an experiment together to discover the laws by which the chemical effects of the pile are produced. These experiments continued until the end of 1802, when I produced the account of them that appeared in the February 1803 number of Gehlen's *Journal der Chemie*. This article contains the foundations of the laws on which the electrochemical theory was later based. The names of the unknown authors did not attract a great deal of attention. But when 4 ½ years later Sir Humphrey Davy applied these laws to produce the most brilliant discoveries that science can show, our envy was surely justified. Davy did not reveal that he knew of our results, but he quoted an experiment that is described in our article and found fault with it. When I presented the French edition of my work on the electrochemical theory and chemical proportions to the Academy of Sciences in Paris in 1819 and was complimented on it by President Vauquelin, he said: "We consider it proper to say by way of redress to you that, if the work of yourself and Hisinger on the chemical effects of the pile had been known to us when Davy was awarded the grand prize, we would have shared it between you and him."[6]

With his usual magnanimity Berzelius commented: "I mention this matter for what it is worth and believe that what happened was the most correct."

Berzelius was to stay in Hisinger's house for ten years. Here he matured not only as a chemist, but also as a cultivated man who was much sought after in the social circles of Stockholm. The period 1810–1820, when Berzelius was in his thirties, was probably his greatest. He had become a professor at the Karolinska Institute and had also acquired other sources of income, releasing him from the need to earn a scanty living as a parish doctor.

Still working with Hisinger, Berzelius had examined the mineral gadolinite, which led to the discovery of a new element, cerium, named after the equally recently discovered minor planet Ceres. Later Berzelius himself discovered the elements thorium and selenium. At about the time of his appointment as professor, Berzelius began his study of fixed chemical proportions. His elucidation of the problems involved, which he undoubtedly regarded as the greatest work of his life, took him until 1818. He then published his table of atomic weights, which was based on quantitative analyses of some 2,000 compounds. With this, Berzelius laid a firm foundation for the future development of chemistry.[7]

In 1812, a year of great political upheaval abroad, Berzelius went to England. He was well received by various members of the Royal Society and was also able to meet Sir Humphry Davy, but this led to discord between them that lasted as long as Davy lived.

It all boiled down to misunderstandings and trifles blown up out of all proportion—a storm in a teacup, in fact. The two scientists were direct opposites in character. Berzelius, the solid student from the provinces, stable possibly to a fault, was the antithesis of the easily offended Davy, who was also inclined since acquiring wealth through marriage to play the *grand seigneur*. Alone in Davy's drawing room, Berzelius contemplated, to quote his words, "the huge quantity of expensive luxuries, little more than warehoused here and giving me the unflattering impression of a nouveau riche with little idea of the meaning of enough."

In Davy's laboratory, however, things were better. The cheering conclusion Berzelius drew from the disorder that he found there was that he had been right in his belief that "a tidy laboratory is the mark of a lazy chemist." At any event, Davy and Berzelius began to find common ground only when they started talking science. Davy, hitherto—in the Swede's opinion—indolent and arrogant, burned with enthusiasm once he got onto his subject. Later, when the rift was undeniable, the ubiquitous Madame de Staël attempted to mediate between the giants, but without lasting success. They met once more, in, of all places, Helsingborg in the south of Sweden. Davy noted coldly in his diary: "Berzelius was fatter than before, when I first saw him—his conversation almost limited to his own subjects."[8]

As professor of chemistry, Berzelius had his laboratory in the former bakery, now long since vanished, of what had once been the palatial city residence of Count Carl Gustaf Wrangel. Conditions there were primitive. In the winter the temperature indoors was often below freezing point. But from there emanated the Berzelian ideas and discoveries that later spread over the world. Among these was the reform of the system of chemical symbols.

The elements were to be denoted by one or two letters from their name in Latin, the first letter of the symbol always being the initial one of the name. Berzelius presented his almost revolutionary ideas on these lines in the *Annals of Philosophy* in 1813.[9] With this he brought order once and for all to a field that had previously been in a hopeless muddle, even if a few years earlier John Dalton had proposed a system of chemical symbols based on circles and strokes. Berzelius' new system caught on quickly and has since been only slightly modified. He himself described it as follows:

"The purpose of the new symbols is not to label laboratory vessels but to serve in facilitating the expression of chemical proportions and to enable us without long circumlocutions to state the relative volumes of the different constituents of every chemical compound. When we have determined the weights of the elementary volumes, we will be able with the aid of these symbols to express the numerical result of an analysis in a manner as simple and easily read as the algebraic formulae used in mechanics."[10]

Only in Britain was there a persistent and determined resistance to what was referred to these as "Berzelius's horrible system." After crossing swords with Davy, Berzelius had never really been in the good books of the chemical establishment in Britain. A commission was set up to devise a system of chemical symbols, but the only result was that nobody could conceive a better system than the one invented by Berzelius—and there the matter rested.

Berzelius was a pure scientist; industrial applications and economics were of little interest to him, even if he was always correct in his affairs in the latter field. He became part-owner of a chemical factory at Gripsholm, whose products included vinegar, soap, and sulphuric acid. While supervising production there in 1817, he discovered the new element selenium—which later acquired practical importance in several different fields, including electrical engineering. Here Berzelius was to work closely with the former wholesale merchant and later research chemist Carl Palmstedt, the true father of Chalmers Technical Institute in Gothenburg. A lifelong friendship was forged between them, and they supported each other in several respects.

The correspondence between Berzelius and Palmstedt is a literary revelation in its outspoken joviality. When the Swedish Academy (of literature and language) made Berzelius a member, it was for the stylistic excellence of his scientific writings. His wonderful letters had not yet become public property; their style is very modern, with a highly expressive, even racy, vocabulary.

While visiting Germany and Austria in 1822, Berzelius called on Johann Wolfgang von Goethe, who was then at Eger. The meeting of these two great minds was initially a little stiff, but when Berzelius demonstrated the technique of blowpipe analysis to Goethe, this aroused the latter's enthusiasm, and he would hardly allow his guest to leave.

Jacob Berzelius (1779–1848) had become a world-famous scientist by the time this portrait was painted in 1843. He was a member of most Swedish and eighty-one foreign scientific academies and societies.

"Goethe is a man of 72," wrote Berzelius, "of medium build, sturdy and thick-set, still without a grey hair, which makes him look like a respectable juryman, and his whole mien was that of a well-dressed, venerable, old-fashioned inspector. He is taciturn rather than talkative, expresses himself simply but not peremptorily and is even more of a true philosopher in his manner than in his writings."[11]

The Regenerator of the Academy

In 1818 and 1819, Berzelius made a journey that included a short sojourn in England and a longer one in Paris. In Paris he met all the prominent chemists with whom he had been corresponding for years. At the end of November 1818, he received a letter from Palmstedt, expressing the heartiest congratulations. Berzelius had been unanimously elected permanent secretary of the Academy of Sciences. This was the start of a new epoch in his life—and, indeed, in the history of the Academy. Not only was Berzelius' period as secretary a happy time for all concerned, it also did much for the international standing of Swedish science. For three decades he was to occupy and develop the central position that the secretaryship of the Academy of Sciences represented.[12]

Berzelius' earlier dealings with the Academy had not been altogether successful. In his youth, for incomprehensible reasons, he had not managed to have his writings published in the Academy's Proceedings. In the election of 1811 to choose a secretary to succeed Jöns Svanberg, he had been defeated by Olof Swartz. The Academy's stock had fallen considerably during recent decades—some said it consisted of "the patrons" and "the patronized"—but Berzelius now breathed new vitality into both its structure and its program of work. The statutes were revised, and the division into sections was modernized; members were prodded into activity, and an annual ceremonial meeting was introduced. He took particular pains with the library, which had previously been neglected or mismanaged. Exchange arrangements were instituted with learned societies abroad. Berzelius also started the Academy's large collection of manuscripts, which was to be of great importance to the future study of the history of science in Sweden.

Another important step taken by Berzelius on the Academy's

account was the acquisition of the big Westman House between Drottninggatan and Adolf Fredrik's Church. This was sorely needed. The Academy's natural history collections had outgrown their home, and activities had expanded in every way.

Berzelius' energies and administrative talent were sufficient for many demanding tasks. The best known of them is his participation in the Government Commission on Education of 1825, as a member of which he fought with great vigor for the place of science in the education of the country's youth. In the struggle between Romanticism and natural philosophy on the one hand, and a critical turn of mind and empirical science on the other, the sides had dug themselves into entrenched positions.

Later Berzelius faced an even more uncompromising battle. The Karolinska Institute was put forward as a desirable alternative to the outmoded medical instruction provided at Uppsala. Berzelius was regarded as the leading representative of the new medicine, and he said what he thought about Romantic natural philosophy. In these discussions the idea was even raised of moving the Uppsala medical faculty lock, stock, and barrel to Stockholm and the Karolinska Institute. But from behind the barricades at Uppsala, stepped the professor of medicine, Israel Hwasser, with his pamphlet *Om Carolinska institutet*,[13] which was aimed directly at Berzelius and is justly known as a remarkable document.

> "Emotional, magnificent in his way," says Sten Lindroth, "using his considerable resources to the full, Hwasser floats off into impenetrable speculations through which surely none of his contemporaries can have been able to find a path. The question of the status of the Karolinska Institute becomes an aspect of the struggle between God and the Devil, between Good and Evil. Totally captivated by the mystical-Platonic ideal of Romanticism, Hwasser regarded the projected removal of the Uppsala faculty to Stockholm as a defection from the eternal order. If that happened, it would mean that medicine lost its divine majesty and was reduced to a simple handicraft. And this, believed Hwasser, would imply the triumph of the greatest evil of the day, materialism or empiricism, the detestable way of thinking that built its world on casual external observation."[14]

In saying this he was pointing his finger at Berzelius, more than anyone else. In private Hwasser mentioned him by name. Berzelius was assuredly the systematic, forceful leader of the materialists in Sweden. But now he had at last found an opponent worthy of him in vigor and independence, with a capacity for righteous wrath to match Berzelius' own. And Sten Lindroth concludes: "Two of the most powerful spirits of the age came face to face." Had the antagonists been able to see 150 years into the future, they would surely have been surprised. It is a long time since anyone has questioned the independent and separate existence of either the Karolinska Institute or the medical faculty at Uppsala.

The Academy in an International Context

Berzelius was to play an international role in science, not only as a chemist, but also as secretary of the Academy and as a publicist. This was by no means inevitable. Some background information may be useful in putting the position of Swedish science in an international context and in setting the scene for an examination of Berzelius' contribution.

When in 1970 the celebrated Swedish physicist Hannes Alfvén received the Nobel Prize, the award was as usual commented on at length in the periodical *Science.* The author of the article noted that Alfvén had been "compelled" during the earlier part of his career to publish his pioneering work in "obscure publications" such as the Proceedings of the Royal Swedish Academy of Sciences *(Vetenskapsakademiens Handlingar).*[15] Although, semantically, the English word derived from the Latin *obscurus* does not have the full range of disapproving connotations of its Swedish counterpart, many in Sweden raised their eyebrows at its use in this connection. This reaction was to a considerable degree due to the fact that for nearly 250 years the Proceedings of the Academy have played a by no means insignificant role in the international exchange of scientific information. There has sometimes been discussion of the strange phenomenon that during the seventeenth and eighteenth centuries, when communications were slow and often dangerous, a unitary European sphere of scientific communication existed, based largely on the use of Latin.

By contrast, the time that has passed since then, for all its technological innovations, has seen a markedly divisive trend. Such phenomena as, for example, cover-to-cover translations would have been inconceivable in the old world. The march towards Babel was led by the Royal Society in London, which began to publish its *Transactions* and other writings in the vernacular in 1660. Shortly afterwards, in 1666, the Académie Royale des Sciences in Paris began to publish its communications in French, and before long Italian and German learned societies added to the confusion of tongues. When in 1739 the Academy of Sciences decided that its publications were to be written in Swedish, it certainly extended its Swedish readership—the quarterly issues of the Proceedings, with their useful papers on various subjects, were in wide demand—but it could also scarcely have failed to realize that the international impact of the Proceedings would be, to say the least, reduced. The great international libraries, however, recognize no language barriers. As early as 1750, the Academy reached an agreement with the Royal Society in London for a full exchange of each other's publications.[16]

To the Academy it was a prime concern to have its work known internationally. By the early 1740s, the Academy was examining the possibility of also publishing its Proceedings in German or French. But steps had already been taken abroad, without any initiative from the Academy itself, to ensure that its writers became known. Moreover, foreign interest in the Academy's work was by no means confined to the output of such international celebrities as Linnaeus, Celsius, Wargentin, and Scheele. The Academy's earliest publications dealt largely with medicine, technology, and political economy. These writings were widely translated and reported, even though many of the authors were unknown outside Sweden. Not surprisingly, German was the language that at first accounted for the bulk of this activity. Until 1814 a highly civilized German population lived under Swedish rule. The soil for Swedish-German cultural relations was particularly fertile in Swedish Pomerania and adjoining parts of northern Germany, and many translations of works by the Academy's authors were published there. The German translations also formed a natural bridge between the Academy's Proceedings and other major languages. Many of the Academy's works appeared in German a few years after the publication of the Swedish originals—although often as much as five years or more later. After a few

To the left is the Westman House, the home of the Academy from 1828 until 1915. Here Secretary Berzelius is receiving King Carl XIV Johan.

more years had elapsed, they would be translated again into other languages.

In the eighteenth century, it was evidently not as vital to be bang up-to-date as it is nowadays. It has been established that translations or abstracts of Academy publications appeared in German, French, English, Latin, Dano-Norwegian, Dutch, and Russian up to the middle of the nineteenth century. After that the Academy itself began to publish its Proceedings, which at that time were in the form of long scientific monographs. Here I will mention only a few important series of translations, such as that published by the eminent German mathematician Abraham Gotthelf Kästner (1719–1800) under the title of *Der Königl. Schwedischen Akademie Wissenschaften Abhandlungen aus der Naturlehre, Haushaltungskunst und Mechanik . . . Aus dem Schwedischen übersetzt.* This series appeared until 1792, when Kästner discontinued it, due either to advancing years or to difficulty in finding competent assistants with a knowledge of Swedish. Among French translations and summaries, particular interest attaches to the impressive volume published in 1772 under the title of *Mémoires de l'Académie royale des sciences de Stockholm concernant l'histoire naturelle, la physique, la médecine, l'anatomie, la chymie, l'oeconomie, les arts etc.* The translator was L.F.G. de Keralio (1731–1795), who had a certain command of Swedish and was an officer and instructor at the Ecole Royale Militaire in Paris. This compendium, which consists of a summary of the first 29 annual volumes of the Academy's Proceedings, is arranged systematically with separate sections for botany, physics, astronomy, chemistry, medicine, political economy, commerce, agriculture, and architecture. The collection is furnished with a detailed index of subjects.

In Holland, too, there was interest in translating the more noteworthy papers published by the scientific academies of other countries. This interest led to the publication of two series of extracts. The first was published by F. Houttuyn in ten volumes under the title *Uitgezogte Verhandelingen uit de nieuwste Werken van de societeiten der Wetenschappen in Europa* (Amsterdam, 1757–1765). This series contains many Swedish contributions, especially by Linnaeus and the celebrated entomologist Charles de Geer. The latter was a corresponding member of the Académie Royale des Sciences in Paris, which published several of his articles. These were later translated into Dutch. The second collection of articles of the Swedish Academy of Sciences to be published in Dutch was compiled by the chief officer of health at The Hague, J.B. Sandifort, who arranged for the translation of many of the Academy's medical contributions for the period 1739–1772 and printed them in four volumes entitled *Geneeskundige Verhandelingen aan de koninglijke sweedsche academie medegeedeeld* (Leiden, 1775–1778).

Parts of the Academy's Proceedings from the years 1739–1752 were translated into Latin and published in Venice in two volumes entitled *Analecta transalpina* (1762); altogether they contained 184 papers. Two Russian scientific journals also contained material from the Academy's Proceedings. One was *Trudy vol'nago ekonomičeskago obestva,* in which items began to appear in 1765, while material was also published later in *Akademiščeskča Izvestija,* the journal of the Imperial Academy of Sciences in St. Petersburg. Single papers by Swedish scientists, published either in the Academy's Proceedings or under the aegis of the Society of Sciences at Uppsala, were also translated into English and appeared in the publications of English learned societies.

Many publications of the Academy, in addition to the contents of the Proceedings, were also translated into French and German. I have deliberately dealt at some length with the position of the Academy in an international context in its early years, as reflected in the interest shown abroad in its publications. Such a reminder may be appropriate, since an otherwise interesting and informative work by David A. Kronick gives no information whatever on Swedish contributions in the domain of science.[17] One may perhaps surmise that the author has shied away from a rather inaccessible language, but the many translations might have been noted, especially as a couple of his chapters are devoted to a lengthy consideration of *The Abstract Journal* and *The Review Journal,* journals that in fact cover publications of very uneven quality and a more or less ephemeral nature and include mathematical and scientific works of little note.

The Annual Reports

On December 20, 1820, the Academy of Sciences adopted new basic rules. The work of revision had been preceded by heated discussion—one member who could not gain a hearing for his ideas resigned from the Academy. As the new secretary, Berzelius, now 40 years of age and recently ennobled, made a number of important contributions. He had spent the academic year 1818–1819 in Paris and had attended a ceremonial meeting of the Académie des Sciences, in which Cuvier delivered one of his reports, so much admired by his contemporaries, on the progress made by natural scientists within the academy's field of activities. Berzelius then conceived the idea of introducing something similar in Sweden as part of his plan to make the Academy of Sciences more efficient and useful after his return. But the Académie, serving as it did a country of 30 million inhabitants, could be self-sufficient, whereas Sweden was at that time in many respects an underdeveloped country, with a population not even one-tenth that of France, and its Academy of Sciences could not be expected to dazzle the world with its discoveries except on rare occasions. At the beginning of the nineteenth century, that is to say during the Napoleonic wars, scientific research and information existed under conditions that are almost inconceivable to us today. Swedish censorship watched Argus-like over everything that was printed in Sweden and kept an equally jealous eye on foreign literature brought into the country. During his time as a physician and a young professor in Stockholm, Berzelius, together with a friend who shared his thirst for knowledge, wanted to form a reading circle that would subscribe to foreign journals in physics, chemistry, anatomy, and medicine. For this it was necessary to obtain permission from the Office of the Court Chancellor, headed by the all-powerful C.B. Zibet.[18] Zibet, a zealous instrument of Gustaf IV Adolf's suppression of free speech, rejected the application out of hand. He pointed out that these young men had finished their studies at the university and that the government expected them to use the knowledge they had acquired there and not to stuff their heads with new-fangled, unnecessary, and expensive knowledge that came from abroad. It was several years before Berzelius succeeded by other means in bringing Gilbert's *Annalen der Physik* and Gehlen's *Journal der Chemie* to Stockholm.

One of the articles in the statutes adopted by the Academy in 1820 provided for a ceremonial meeting to be held on March 31 every year. As a result of Berzelius's initiative, arrangements were now made for annual reports on progress in the natural sciences. These reports were to be clear and easy to understand and were to embrace all accessions of knowledge that the reporter considered worth presenting. Reference was to be made to all the literature dealing with science and related subjects. Medicine and

agriculture were not included, as these subjects came under the
Swedish Medical Association and the Royal Academy of Agriculture, respectively. Berzelius was, of course, aware that he alone
could not be responsible for all these wide-ranging reports with
their references to printed literature. But he relied on the senior
officials of the Swedish Museum of Natural History, which came
under the Academy, and especially on the head of the zoological
collections, who was to report on works in comparative anatomy
and zoology. He was similarly confident of the ability of the head
of Stockholm Observatory, who also belonged to the Academy, to
carry out the planned surveys in his subject. The Bergius Professor
(named after the botanist and benefactor of the Academy Peter
Jonas Bergius), who was in charge of the Academy's botanical
garden, was to be responsible for reports on all branches of botany.[19]

In addition to chemistry, Berzelius made himself responsible
for physics, mineralogy, and geology, although, as he later wrote,
he did not consider himself more than a dilettante in these three
subjects. As regards technology, Berzelius doubted whether he
would be able to induce the extremely self-willed and obstinate
Gustaf Magnus Schwartz to produce annual reports on that subject. Berzelius' fears were confirmed—Schwartz dragged his feet
over the first few reports—and not until Gustaf Erik Pasch became
head of the Academy's Institute of Physics did the reports on
technology reach a satisfactory standard. Pasch also relieved Berzelius of the task of producing reports on physics.

At first the annual reports were printed with continuous pagination, i.e., as parts of a single work. However, the principles of
editing had soon to be altered, as the various authors seldom had
their reports ready at the same time. The more diligent reporters
might therefore have to wait an unconscionably long time before
seeing their efforts in print. Consequently a change was made to
individual pagination for each section, enabling the reports on
different subjects to be sold separately in the bookshops; thus a
researcher in a certain field did not have to pay for reports in
which he had no interest. After some time these individual reports
were furnished with subject and author indexes, which, of course,
made them much easier to use.

On March, 31 1821, the first ceremonial meeting in the history
of the Academy took place. The occasion was celebrated with
great pomp and circumstance in the then hall of ceremonies of the

Academy in the Lefebure Mansion at Lilla Nygatan 13, Stockholm. Berzelius presented his annual reports in an address lasting only half an hour—a considerable feat of condensation, since in printed form the reports filled 160 pages.

While Berzelius' annual reports clearly constitute the prototype for subsequent scientific information publications, they also have unique features. Above all one is struck by the literary mode of presentation. Berzelius was eager for the necessarily compressed mass of information to be presented in such a way as to make connected reading, so that the reports would not be used only for reference purposes. He himself wrote on one occasion that part of his intention in arranging for the annual reports was that they should stimulate the curiosity of the public and arouse a desire to know more, and that, if possible, they should be so clear that even a reader who was not a trained scientist would not only become interested, but also be able to understand most of the contents.

Berzelius himself found the task of compiling his annual reports extremely interesting and stimulating. He freely acknowledged that the self-imposed duty of going through innumerable foreign publications in order to be able to produce a satisfactory annual report on chemistry "enabled me to achieve a level of knowledge which otherwise I should never have attained."[20]

Another distinguishing feature of his annual reports was that he not only summarized the publications, but often added his own reflections. For a scientist with such a well-developed critical sense, it was almost impossible to refrain from challenging statements that he considered fallacious or obscure. Berzelius also had a more pedagogic rationale for this policy. In his first report on the advances in chemistry, he stated that the series was intended primarily to be used by the young student. Consequently he took as his starting point his own textbook on chemistry, the annual reports being composed so that they could be treated as supplements to that work. It was thus necessary to subject all literature reported to critical scrutiny, "to point out probable inaccuracies and misleading statements in order thereby to accustom the young student not to have blind faith in everything that was printed in the scientific literature."[21]

The criticisms contained in the reports meant, of course, that Berzelius made enemies among those who felt the lash of his pen, but—particularly in view of the author's great authority—the reports were also received with tense interest and read attentively by the entire scientific world. Berzelius himself was but little perturbed by the storm his attacks aroused. For a man of his upright and uncompromising nature, scientific truth as he conceived it was all that mattered. There are many examples of this, but here it will suffice to mention one.

In 1825 the eminent Scottish chemist Thomas Thomson had published a work entitled *An Attempt to Establish the First Principles of Chemistry by Experiment.* This book was ambitious in conception, the author regarding it as the final fruit of his long labors, especially on atomic theory, and the crown on his scientific career. Berzelius, in his annual report for 1826, demolished Thomson's book point by point and made the following devastating comment:

> "This work is among the few from which science has derived no gain. Much of the experimental content, including even the basic experiments, appears to have been elaborated only at the desk, and the greatest courtesy that his contemporaries can show the author is to consider this work as never having been published."[22]

It is relevant to mention that in a printed dedication on the endpaper Thomson had included Berzelius' name. "It is an easy way out," wrote Berzelius, "to purchase praise with praise. I have never resorted to it; instead in judging the works of others I have always served the best interests of science as I have understood them."[23]

It is scarcely surprising that the annual reports presented by Berzelius and his colleagues to the Academy attracted much attention. The well-known Swedish dignitary Count Hans Gabriel Trolle-Wachtmeister, who was also a distinguished chemist, wrote to Berzelius: "Thanks to the orientation you have given them, the Academy's annual reports will introduce a new epoch in our scientific life."[24]

Just as the Academy's Proceedings had reached a wide circle of readers, at least in German and French-speaking areas, during the eighteenth century, its annual reports on the advances of science were also studied with interest. In Germany the task of making the reports accessible to German-speaking readers was at first shouldered by the celebrated Christian Gottlob Gmelin of

Tübingen, and the first volume of the reports soon appeared in a German translation. Sometimes, however, the interest and acclaim with which the reports were received were mingled with protests and expressions of indignation. According to Berzelius, one German chemist had mistaken a certain copper-bearing earth for a new metal. This was mentioned, more or less in passing, by Berzelius in his first annual report. The chemist in question, understandably nettled, retaliated by declaring in a review that Gmelin had done no service to science or the German public by translating this "foreign product."[25]

A circumstance of great importance to the dissemination of Berzelius' annual reports was his acquaintance with the young German chemist Friedrich Wöhler. The latter lived until 1882, by which time chemical science had entered a new phase, and Wöhler himself was regarded as the grand old man of chemistry, not only in Germany. In his youth in the 1820s, Wöhler had worked in Berzelius' laboratory, where he enjoyed the master's favor and confidence. During his studies in Stockholm, Wöhler soon learnt Swedish. He set to work on a translation into German of Berzelius' *Lärbok i kemien,* a work that was constantly reprinted in new and expanded editions.[26] So long as Berzelius lived, Wöhler saw to it that his ideas and results reached new linguistic areas and were made available to at least the German-speaking public. With the annual reports of 1825, Wöhler also took over the translation of the volumes, Berzelius having become increasingly dissatisfied both with Gmelin and with the printing house that produced *Jahresbericht der Fortschritte.*

But Wöhler, too, was to be placed in an invidious position by Berzelius' critical treatment in the annual reports of certain writings that in his opinion were of inferior quality. In Poggendorff's *Annalen,* a Berlin physicist, G.F. Pohl, had published a paper on "Der Process der galvanischen Kette." In the annual report for 1826, Berzelius had dismissed this work with the following words:

"If meaningless expressions and absurd comparisons could make up for deficiencies of matter and content, the title of this work would be promising. As it is, one can only regret that there are those who do not realize that within the sciences there applies the well-known saying *tamen est laudanda voluntas,* the essence of which may be con-

veyed by the rather free translation 'the will to do something is not enough'."[27]

While Wöhler was engaged in the task of translation, he received
a visit from a common friend of his and Berzelius', T.J. Seebeck.
Seebeck expressed the opinion that Pohl had been too harshly
treated and advised Wöhler to omit the offending passage from
the German edition and to replace it with dashes. Wöhler forwarded this suggestion to Berzelius and received the following
reply:

> "Convey my greetings to Dr. Seebeck and say that I hope
> I may be allowed to answer for myself for what I write and
> that there is no need for my friends to adopt the role of
> guardians and to leave things out and fill up the space with
> dashes. The proposal is absurd; I give you full authority to
> remove the whole account of Pohl's idiocy, but if it stays
> then my verdict on that rubbish also has a right to be
> printed. To mention it without criticism would be a crime
> against the sanctity of science. To pass over it altogether
> in silence as an expression of contempt I am, however,
> prepared to allow."[28]

The outcome was that the scathing criticism of Pohl's work did
after all appear in *Jahresbericht der Fortschritte.* Pohl swore ven-
geance. In his capacity as a natural philosopher, he was already in
dispute with Berzelius, and he therefore took the opportunity in
Jahrbücher für wissenschaftliche Kritik to write in disparaging
terms of Berzelius' textbook on chemistry. Berzelius' supporters
and others who opposed the natural-philosophy school within the
natural sciences accused Pohl of having "for reasons of personal
revenge criticized a work for the evaluation of which he lacked
the necessary scientific equipment."[29] With these examples of
odium academicum, we can leave Wöhler and his translations of
the Academy's annual reports.

Philippe Plantamour was a young Swiss who, after a period
with the celebrated Justus von Liebig of Giessen in 1838, had the
unusual privilege of being accepted as an assistant in Berzelius'
own laboratory. He quickly acquired a knowledge of Swedish and
was soon entrusted with everything that the master wished to
have translated into French. Thus Plantamour also took on the

RAPPORT ANNUEL

SUR LES

PROGRÈS DES SCIENCES

PHYSIQUES ET CHIMIQUES

PRÉSENTÉ LE 31 MARS 1840

A L'ACADÉMIE ROYALE DES SCIENCES DE STOCKHOLM,

PAR

J. BERZELIUS,

Secrétaire perpétuel.

TRADUIT DU SUÉDOIS, SOUS LES YEUX DE L'AUTEUR,

PAR M. PLANTAMOUR.

⸻⸺◦✦✦◦⸺⸻

PARIS,

FORTIN, MASSON ET Cⁱᵉ, LIBRAIRES–ÉDITEURS,
4, PLACE DE L'ÉCOLE-DE-MÉDECINE.
—
1841.

The title page of the French translation of Berzelius' annual reports.

formidable task of translating the annual reports, beginning with the volume of 1840. The translations were published in Paris under the title *Rapport annuel par les progrès des sciences physiques et chimiques presenté le 31 mars . . . à l'Académie Royale des sciences de Stockholm, par J. Berzelius. Traduit du suédois sous les yeux de l'auteur, 1841-49,* (8 vols.). These annual reports in French, generally containing about 500 pages each, appeared as long as Berzelius lived, that is to say until 1848.

It is notable that the annual reports were never published in English, probably because during the 1830s Berzelius became the target of criticism from British chemists and was far from popular among them. We have seen that Berzelius visited England in 1812, when he met British colleagues, including Sir Humphrey Davy. His relations with the latter deteriorated after Berzelius in his forthright manner had criticized a work by Davy. I have dealt earlier with the controversy with Thomas Thomson. The new scheme of chemical formulation introduced by Berzelius did not—as has been mentioned—readily find favor in Britain, where many recoiled from this radical departure from the old familiar system. Gradually, however, British chemists began to accept and use it.

Berzelius' annual reports had no direct successors. After his time the indexes of the leading chemical journals took over most of the task of mediating international scientific information. Reports and abstracts as we know them today are a product of the first decades of the present century. Berzelius' attempt year by year to epitomize and evaluate the scientific output of his fellow chemists is in itself extremely impressive. To undertake anything on this scale at the present time would be unthinkable without a huge staff. It is, moreover, highly unlikely that any scientific institution or publishing house would wish to take on the responsibility of publicly judging the great names of the world in, for example, chemistry with any claim to be uttering abiding truths.

Concluding Remarks

Only some of the aspects of Berzelius' work for the Academy have been mentioned here. His informational activities probably meant more than is usually appreciated, particularly in an international perspective.

But it is also appropriate finally to recall the breadth and diversity of Berzelius' work. He was a busy traveller, by the standards of the time, and he visited his leading colleagues in their home towns. He also met outstanding cultural figures, such as Goethe and Humboldt. When he visited the Great Scientific Congress in Hamburg in 1830, his greatest benefit was making the acquaintance of Justus von Liebig. The two men took to each other, and Liebig planned to spend a couple of months in Stockholm. However, this proved impossible to arrange, which was unfortunate, because it might have forestalled the dispute that arose between them later. Berzelius' talented pupils have been mentioned, most important among them being Heinrich Rose, Eilhard Mitscherlich, and Friedrich Wöhler. It was natural that Berzelius' pride was wounded when Humphrey Davy spoke slightingly of Wöhler; this Berzelius could not accept.

Berzelius' acknowledged excellence as a correspondent also shows another side, his humanity and loyalty. His closest friends were kept regularly entertained with long letters, recounting what he had been doing and readily answering questions on chemical subjects. His relationship with Count Hans Gabriel Trolle-Wachtmeister is interesting and touching.[30] Trolle-Wachtmeister was appointed Attorney General at the age of 27 in 1809, after the coup that deposed Gustaf IV Adolf. The young official had a passion for chemical experimentation, and in his spare time studied with Berzelius. In 1817 he resigned from office to devote himself to his estate in Skåne in the south of Sweden. He immediately converted an old henhouse into a chemical laboratory and carried on with his hobby. He remained in frequent contact with Berzelius, from whom he received support and encouragement. In time he became a successful experimenter. Berzelius sometimes met his German colleagues on Trolle-Wachtmeister's estate, which lay about halfway between them.

It was not only in Germany that congresses of scientists were arranged. In Scandinavia, too, there was a wish to use this form of contact for international exchange. Berzelius was actually against the idea as such. He felt that Scandinavian meetings became too parochial, and that it was more important to reach out to the wider world beyond. He believed it better for young scientists to have the opportunity of attending the really big gatherings in Germany. But he nevertheless turned out dutifully and took an active part in several Scandinavian meetings: in 1840 and 1847 in Copenhagen, in 1842 in Stockholm, and in 1844 in Kristiania (Oslo). It must also be remembered that by this time Berzelius was beginning to feel tired, worn out, and sometimes ill. He was a bachelor for most of his life, dividing himself between the Academy and the laboratory in about equal proportions. But in 1835 he surprised everyone by getting married; he was then 56, and his young bride, Betty Poppius, was 24!

In his final years, infirmity and illness began to make themselves increasingly felt, even though Berzelius continued to work doggedly as long as he had any strength left. The unhealthy laboratories of the day were not only cold and draughty, but must also have given rise to poisoning in various forms. Berzelius not infrequently overexerted himself. He spent his last days in a wheelchair, and died in August 1848. In the development of science and the Academy of Sciences in Sweden, Berzelius' contribution is impossible to exaggerate.

Notes

1. Arne Holmberg, *Berzelius-porträtt: Illustrerad beskrivning.—Portraits du chimiste suedois J.J. Berzelius: Illustrations et descriptions* (Stockholm, 1939).
2. J. Erik Jorpes, *Jac. Berzelius: His Life and Work*, Bidrag till Kungl. Vetenskapsakademiens historia, 7 (Stockholm, 1966); new ed. (Berkeley, 1970); H.G. Söderbaum, *Jac. Berzelius: Levnadsteckning*, 3 vols., utg. av Kungl. Svenska Vetenskapsakademien (Stockholm, 1929–1931); H.G. Söderbaum, *Berzelius' Werden und Wachsen, 1779–1821*, Monographien aus der Geschichte der Chemie, herausg. von Georg W.A. Kahlbaum, Vol. 3 (Leipzig, 1899). A short but brilliant biography is Sten Lindroth, "Berzelius och hans tid," in *Vetenskapssocieteten i Lund: Årsbok* 1964, 15–40; also available in French translation, "Berzelius et son temps," in Lindroth, *Les chemins du savoir en Suède*, tr. J.F. Battail (Dordrecht, 1988), 209–232.
3. Berzelius' autobiography was published as *Själfbiografiska anteckningar*, utg. av Kungl. Svenska Vetenskapsakademien genom H.G. Söderbaum (Stockholm, 1901), 175.
4. *Ibid.*, 177.
5. Depository of Berzelius' manuscripts, correspondence, and laboratory equipment, etc. is now in the Center for History of Science at

146   Science in Sweden

the Academy. Cf. Arne Holmberg, *Bibliografi över J.J. Berzelius,* II (Stockholm, 1936); also with French translation.

6. Berzelius, *Självbiografiska anteckningar,* 32.

7. Gunnar Eriksson, "Berzelius och atomteorin: Den idéhistoriska bakgrunden," *Lychnos* 1965–1966, 1–37; Anders Lundgren, *Berzelius och den kemiska atomteorin* (Uppsala, 1979), with an English summary: "Berzelius and the chemical atom theory."

8. John Davy, *Memoirs of the Life of Sir Humphrey Davy* (London, 1936), 2, 209. A modern description concerning the controversies between Berzelius and Davy is given by Harold Hartley, *Humphry Davy* (London, 1966).

9. Berzelius, "Essay on the Cause of Chemical Proportions and on Some Circumstances Relating to Them: Together with a Short and Easy Method of Expressing Them," *Annals of Philosophy,* Vol. 1 (1813), 443–454, Vol. 2 (1814), 51–62.

10. *Annals of Philosophy,* Vol. 3 (1814), 51.

11. Jan Trofast, ed., *Brevväxlingen mellan Jöns Jacob Berzelius och Carl Palmstedt,* utg. av Kungl. Svenska Vetenskapsakademien (Stockholm, 1979), Vol. 1, 209–210. Quotation from *Självbiografi.*

12. Söderbaum, *Jac. Berzelius,* 2–3.

13. Israel Hwasser, *Om Carolinska Institutet* (Stockholm, 1819), 2nd ed. with the author's name (Uppsala, 1860). Cf. Sven-Eric Liedman, *Israel Hwasser,* Lychnos-Bibliotek 27 (Uppsala, 1971), 149–174.

14. Lindroth, *Les chemins du savoir en Suède,* 228.

15. Cf. Svante Lindqvist, "Ett experiment år 1744 rörande norrskenets natur," in *Kunskapens trädgårdar: Om institutioner och institutionaliseringar i vetenskapen och livet,* eds. Gunnar Eriksson, et. al. (Stockholm, 1988), 40–77.

16. Lindroth, *KVA Historia,* 1:1, 168–216. A. Holmberg, Kungl. Vetenskapsakademiens äldre skrifter i utländska översättningar och referat. Mit deutscher Zusammenfassung, *KVAÅ* 1939, bilaga.

17. David A. Kronick, *A History of Scientific and Technical Periodicals: The Origins and Development of the Scientific and Technological Press 1665–1790* (New York, 1962).

18. Söderbaum, 1, 159 f.

19. Einar Lönnberg, et al., *Naturhistoriska riksmuseets historia: Dess uppkomst och utveckling.* (Stockholm, 1916). V. B. Wittrock, "Några bidrag till Bergianska stiftelsens historia," *Acta Horti Bergiani.* Vol. 1 (1890). J. Ramberg "Stockholm observations 200 år," *Populär astronomisk tidskrift* 1953, 96–121, and *KVAÅ* 1959, 307–341.

20. Söderbaum, 2, 238–239.

21. *Årsberättelser* 1921, 19 ff.

22. Årsberättelser 1926, 82; cf. 183. See also Jack B. Morrell, "Thomas

Thomson," in *Dictionary of Scientific Biography*, 13 (New York, 1976).

23. Söderbaum 2, 527–529.
24. Quotation from Trolle-Wachtmeister to Berzelius 22/3 1821. Cf. Söderbaum 2, 238–239.
25. *Årsberättelser* 1821, introduction.
26. The German edition: *Lehrbuch der Chemie*, tr. F. Wöhler, 4 vols. (Dresden, 1825–1831).
27. *Årsberättelser* 1826.
28. Berzelius' letter to Wöhler 2/2 1827.
29. Söderbaum 2, 438–439.
30. Jan Trofast, *Excellensen och Berzelius* (Stockholm, 1988), gives the full story of friendship and correspondence between the two.

GUNNAR BROBERG

The Swedish Museum
of Natural History

OF all the undertakings of the Academy of Sciences, the Swedish
Museum of Natural History was long one of the most important.
In terms of finance, it probably cost more than all the rest together. It also attracted more notice from outside the Academy
than any other activity. This was as it should be: a museum is there
to reach a public. But it also exists for research and documentation.

The following chapter sets out to depict the growth of the
Swedish Museum of Natural History into an imposing colossus of
stone on the Frescati plain near Stockholm, the scene of a lengthy
chapter in the history of Swedish science, a shrine for the lover of
natural history, and a climatically controlled repository of, nowadays, some 15 million objects. The full story cannot be told: space
is lacking and considerable gaps remain in the basic research. The
events of more recent years are still stored in files or in the hearts
of individual officials. The material in the files (apart from the
natural history collections themselves) is in any case perhaps not
as impressive as might be expected. We may particularly feel the
lack of material illustrating the interface between museum and
man. Nonetheless, this is one of the themes that we have chosen
to deal with. Other matters considered in outline are the organization and finance of the Museum, the growth of scientific research,
and the increasing specialization.

From Cabinet of Curiosities to Seat of Research

Museums of this kind have their origins in the old cabinets of curiosities, but also in the botanical gardens, pharmacies, and menageries. "Giants' bones" (often parts of whales' skeletons) on public display in the churches put these institutions, too, among the predecessors of the museums.[1] They might have links with university libraries and also with royal collectors or private enthusiasts. The underlying philosophy, more or less explicit, was a kind of ambition to reveal the world in both its diversity and essence. The hope of collecting everything in the world in microcosm in a display case, the old *Kunstschrank,* had obvious religious overtones: see the multiplicity, the beauty, the dexterity—but see also the God-given order, the hierarchy, the scale of values! Equally favored by sovereigns and great magnates was the cabinet of curiosities—the small room in which might be accumulated a series of proofs of real or imagined power and breeding.

Such cabinets were also created for more or less educational or instructional purposes; one seventeenth-century Swedish example is provided by Olof Rudbeck's collections in Uppsala.[2] The mixture of art and nature continued into the eighteenth century, when collecting had definitely become the done thing for the cultivated man. His study, or "museum," was filled not only with books, but also with stuffed or preserved animals, with the sheets of his herbarium, and with his mineral samples. Linnaeus' great project, describing nature in its entirety down to the tiniest detail, was extremely important for the growth of modern museology. The previous shapeless accumulation of exotic trivia and freaks of nature lost ground and finally fell into disrepute. During the eighteenth century, collecting also became popular in the most exclusive circles in Sweden, right up to the royal couple Adolf Fredrik and Lovisa Ulrika, as a consequence of both a general interest in the exotic and the rise in the status of science.

The changes in the content and size of museums reflect the changes in science and society. For a long time, for example, private individuals could keep museums that rivaled the national ones. A case in point was the museum of Charles Wilson Peale in Philadelphia (1784–1845), in which curiosa rubbed shoulders with a Linnean layout.[3] In England were the museums of Sir Ashton Lever and William Bullock—the latter more of a showman, who brought Swedish Saami (Lapps) to England in the 1820s as exotica. In Sweden the collections of the Grill family at Söderfors Iron-works in Uppland may be noted, as may those of Gustaf von Pay-kull on the Uppland estate of Wallox Säby. But as the demand for comprehensive coverage grew, it became increasingly unrealistic for the private individual to compete with the national museum.

The eighteenth-century focus on taxonomy and the world inventory became, if anything, stronger in the following century.[4] The museums had taken over the international role of the universities as centers of scholarship. In England colonialism provided a stimulus. Something similar might be said about the origins of the German museum of colonial science in Hamburg. With this the museums became enormous banks for the rich harvests of the natural historian, continuously topped up by travellers and donors. A list may illustrate the increase in knowledge: in 1758 Linnaeus described 564 bird species; in 1760 Brisson counted some 1,500, in 1790 Latham 2,951, in 1812 Illiger 3,779, in 1841 Gray approximately 6,000, in 1871 Gray 11,162, and in 1909 Sharpe 18,939 species. Corresponding with this increase (which is, of course, partly the result of a sub-division of known species) was a growth in the number of museums. The same steeply rising curve is also found for other groups of plants and animals. The plants and animals described by Linnaeus total some 15,000, whereas today the number of known species of insects alone is between 20 and 40 million, an increase primarily attributable to the huge number of insects discovered with the exploration of the rain forests.[5] The British Museum, for example, is estimated to have approximately 55 million specimens.

But while museum biology grew, inherent problems persisted. In some fields taxonomic research was confirmed by the breakthrough of the theory of evolution; in others it was not. Towards the end of the nineteenth century, a struggle developed between traditional taxonomy and laboratory-oriented experimental biology. To many of those engaged in physiological and genetic study, collecting could appear old-fashioned. Nevertheless, the first decade of the twentieth century was a great period for founding museums, particularly in the Teutonic world (the busiest period in England was just after World War I). The museums were concerned to assert their value to research and also to be seen as independent of the universities. The museums also showed a growing degree of specialization. Nevertheless, a museum of natu-

ral history has to be large if it is to have anything to offer. This is
why the museums expanded during a period that otherwise appears to be one of crisis. For the smaller, individual university
museums, times grew harder. Such a climate provided ample opportunity for friction between university and museum zoologists.

The educational duty of the museums became ever more inescapable with the democratization of society. This might involve
longer hours of opening, a more instructive—or sensational—
method of display, and public lectures. World fairs are also worth
mentioning since, they were inclined to show natural history as
well as having cultural themes. Exotic oddities were particularly
popular. Some museums were created to put such exhibitions on
a permanent footing, which, if nothing else, gave the public a taste
for seeing the world as one enormous showcase. In this way the
museums had become both pleasure palaces and scientific institutions.

Collections before the Swedish Museum of Natural History

From its inception the Academy of Sciences possessed natural
history collections of, to put it mildly, a variegated nature, donated
by the members as the occasion arose.[6] Here the old character of
the cabinet of curiosities remained long in evidence: Count Carl
Bonde presented an embalmed thumb, reported to have been cut
off a lake monster several hundred years before. On the mixture
of *objets d'art,* medicine, and natural history there is less to be
said: there were donations of archaeological finds, thunderbolts,
runic staffs, coins, and embryos preserved in alcohol, *praeparata
anatomica;* on the ethnography side, the outstanding item was a
Hottentot embryo in alcohol given by the traveller to the East
Indies Daniel Gottlieb Lange.

But above all representatives of the three kingdoms of nature
flowed into the Academy. A collection of great scientific importance was the primarily entomological one donated by Charles De
Geer's widow in 1778; from Anders Sparrman came large collections of fauna, augmented after the voyage to Senegal in 1788,
from Mine Councillor Nils Psilanderhjelm an outstanding collection of minerals in 1761, and from Peter Jonas Bergius in 1784 a
copious herbarium of some 16,000 species, many of them type
specimens; at about the same time, the Linnean pupil Lars Montin

donated his large plant collection. Material was received from such travellers as Bengt Euphrasén and Samuel Fahlberg in the colony of Saint Barthélemy, and in the 1790s Pastor Nils Collin of Philadelphia gave the skeletal remains of a mastodon, one of the great sensations of the day. In due course, parts of the extremely well-arranged collections of King Adolf Fredrik and Queen Lovisa Ulrika arrived; the Linnean harvest continued. From Swedish businessmen and travellers, came more material, obtained in some cases with great pains and at great financial sacrifice by the individual concerned or by the Academy. It is strange to find at the end of the century one of the four known examples of the bluebuck, previously in the Grill collection, the animal meanwhile having become extinct in the 1790s, also a quagga foal, given by Sparrman and representing a species that has not been seen since the 1870s.

Donating the results of one's labors was a way of buying one-self into a grateful Academy. This does not mean that the generosity did not create problems, for the donations naturally required both space and care. The premises were for many years inadequate, just a couple of rooms in the Lefebur House on Nygatan with the curator's living accommodation alongside. Anders Sparrman, a bachelor whose round-the-world voyage with Cook was his main and virtually unbeatable qualification for the post, became the Academy's first curator in 1779. However, he soon showed that he was not the right man. There was no questioning his specialist competence, but he displayed little ability to keep things in order and maintain contact with his employers. A much more suitable person for the job was Sparrman's deputy, the military doctor Clas Fredrik Hornstedt, who had travelled to the East Indies and was curator for just over a year in 1787–1788. Under Sparrman it sometimes seemed as if the Academy did not really want more material.

The dissatisfaction with Sparrman compelled the Academy to look for a substitute, and the appointment in 1798 of the capable zoologist Conrad Quensel brought a change for the better for a few years. Quensel's death at age 38 in 1806 was particularly unfortunate, but his successor Olof Swartz, with his excellent qualifications in botany, was a good choice. However, when Swartz became the secretary of Academy, his energies were not sufficient for the collections as well, and these began to fall into neglect. A start on cataloguing and restoration was made time and again, only for work to be broken off before the job was finished. We should note that Quensel's period in office saw the appointment of Sweden's first professional taxidermist. This was Anders Berggren, who had learned his art in Paris and then practised it at Wallox Säby. He had a competent colleague in Pehr Gustaf Lindroth, who worked for the Grill family at Söderfors. The results of Berggren's and Lindroth's efforts were soon to be seen side by side in the Museum.

The Academy's museum work during the first hundred years of its existence was erratic and shows little sign of planning. The spirit of Linnaeus hovered over it all, but it was less botany than zoology that dominated, and so it would remain. We know much about how the collections were displayed. Instructions dated 1784 prescribed that cabinets of fish, amphibians, and other creatures in jars were to be arranged in neat symmetry, the glass-fronted cases of corals and zoophytes to be placed for maximum effect, and the smaller molluscs in special glass stands to occupy a position in the center of the floor; in keeping with contemporary practice, stuffed small birds were to be kept in bell jars, while larger birds and mammals were placed wherever suitable. Each class of animals was in theory to be kept together, but aesthetic considerations and the size of the jars had to be considered. It is to be noted that admission to the Academy's exhibitions was free. They were formally announced and advertised in the press in 1784, during Sparrman's period. This is early; Sloane's and Hunter's collections in London may have been opened to the public at about the same time, but the collections in Paris were not put on display until after the French Revolution.

The Paykull Donation and the Museum of Natural History

The Museum's prehistory is far better known than its history proper, which begins in 1819 with Gustaf von Paykull's donation of his collections. These were larger than those in the private possession of any other Swede: approximately 80 mammals, 1,362 boxes of birds, a collection of eight cabinets of insects, etc.; Paykull's speciality was entomology. Somewhat later the collection of Adolf Ulrik Grill, containing about 700 birds and over 80 mammals, was received. Paykull and Grill were outstanding examples of the eighteenth-century private collector, who had now reached his limits, in terms of finance, accommodation, and scientific knowledge. The idea of making a donation of this kind may have come to Paykull from one of a number of sources, perhaps long before from two German experts who visited Sweden to check on its collectors. By 1804 they had reported that Paykull owned 1,200 bird species and that he said that he lacked only 50 of those that were to be found in the National Museum in Paris.[7] The two Germans comment on how well his and the royal collections complement each other—they also remark that the Uppsala University collections could form an excellent focal point: "What treasures for the study of nature unite Sweden!" Transporting the material to Stockholm was no easy matter. How should the large mammals be packed, the delicate butterflies protected and padded, and the birds' nests kept in one piece? One journey required 11 carts, another 14 pairs of horses and 30 men—surely two of the most remarkable transports in Swedish history. From 1820 state grants were received for the care of the collections. In 1828 they were transferred to the building at Adolf Fredrik's church square and were opened to the public in November 1831.

With this, the Swedish Museum of Natural History had been born. The Museum was an offshoot of the eighteenth century's restless delight in collecting, which in principle aspired to complete knowledge, no less, of nature in all its diversity. The Museum—like similar museums in other countries—has to be seen as a kind of encyclopaedia, in which the different rooms are the separate volumes, the shelves are the chapters, and the specimens are the individual articles. The purpose of the new institution is nowhere explicitly recorded. The title "National Museum" (Riksmuseet) was used from its conception in 1819; Paykull's donation was to serve as a "central collection for the natural history treasures of our native country." Paykull himself was interested in the national fauna and had published a three-volume entomological fauna of Sweden (1788–1800). Sven Nilsson, curator for a few years of expansion, was particularly instrumental in strengthening the national emphasis. Nilsson tried several times to give the sportsman a role as a collector "to the benefit of the scientist and the honor of our nation." This produced many donations to the Museum and the splendid but short-lived *Tidskrift för jägare och naturforskare* (1832–1834) as a unifying journal. Nilsson was the leading representative of the hunting tradition at the Museum, although this continued into the next century.[8]

The post of curator of the botany section, which was combined with that of director of the Bergius Garden, was held by Johan

Emanuel Wikström. A taxidermist for the zoological side and a caretaker made up the entire the original staff. The permanent secretary of the Academy, Jacob Berzelius, also had overall responsibility for the Museum. In 1834 a subscription enabled a position to be created for a five-year period for an artist and painter. The post was first filled by the excellent Finnish animal painter Wilhelm von Wright, whose poor health, however, meant that he was often absent convalescing in Bohuslän. Later, to accelerate the publication of Elias Fries' great mycology, which was dragging on year by year, the painter Peter Åkerlund was engaged.

Insight into the activities of the Museum in the 1830s is provided by the letters of Bengt Fredrik Fries (no relation to Elias), curator until his early death in 1838. An exclamation of delight from 1830: "The professor may be sure that there are gold mines here, and that I feel as if I were in paradise—are but life and health granted me, the Museum shall be put in order, and I greatly hope *grow enormously.*" The sheer quantity was surely part of the great fascination to the museologist. Fries presented (1831) a program for gathering new treasures: "We are answerable to the whole scientific world for the way we deal with our own fauna." The foreign interest particularly applies to the natural history of Lapland. A limited company was to be formed for this purpose. An expedition duly set off for Lapland in 1832. "The main thing is to collect Lapland Insects in quantity so that the partners will get some value for their 50 *riksdaler.*" However, the various signatories refrained from realizing their profits in favor of the Academy. Fries intended to go out to the skerries in 1832 to acquire the marine creatures "that we dare not spare for foreigners."

On a trip abroad in 1833, Fries commented on current methods of conservation: "Fredrik Blank [the taxidermist] has nothing to learn here yet, but he might well give lessons to the Berliners." The Museum must be enlarged to hold its own in international competition, "but this requires money, nothing but money." One possibility was to dispatch expeditions, to the Himalayas, for instance.

> "Mr Professor [Berzelius] might possibly object that such a journey may well cost more than the objects obtained therefrom are worth, but so certain am I of a 300 percent profit for the company that if I owned a private fortune and were after a large killing, I would make much money

by putting it into such an enterprise. This may be thought illiberal," said Fries, "but it must be done."

Nature costs money, the Museum is a business. The unflagging Fries later tried (1836) to interest the Academy through its secretary in an expedition to collect material from the Cape, which led in due course and with the support of donors to Johan Wahlberg's many years of fruitful collecting in South Africa. These ended in tragedy when Wahlberg was trampled to death by a rampaging elephant—or possibly murdered by his bearers (1856). His blood-stained diary survives in the cupboards of the Academy to this day.

Nevertheless, after seeing the museums of Berlin, Frankfurt, and Paris, Fries was satisfied with his own in Stockholm. "With all respect to them and with gratitude for what I have learned from their strengths and weaknesses, I must for the time being call them mere depositories, where the one object seems to ask the next 'who art thou and why art thou here?'" As a Swede he insisted on systematics, and on his travels he had new ideas on how a museum should be laid out. "Here in the cradle of the revolution it was inevitable that I should pick up the ideas of the revolution, and as I have no other creatures to try them out on than my quadrupeds and fowls, these will have to become republicans."[9] The reasoning fits an externalist interpretation of science almost too well, but, of course, it contains a lot of rhetoric. The crux is the conflict between diversity and order, between pure collecting and taxonomic treatment.

The main source of the collections was donations—entomological collections, herbaria, lapidaria, individual objects, foreign and native rarities. In 1826 the landowner Alexander Seton donated a pair of large antlers from a bull elk (possibly brought down by Charles XI); the same year also yielded an albino magpie, a frog with five legs, and a whole walrus head from Hammerfest on the Arctic Ocean. The anthropological gifts included the skulls of a Lapp and a New Zealander. In 1820 a live rattlesnake—intended recipient unspecified—arrived. The reports nevertheless show both Sweden and the field of botany as underrepresented. But here as elsewhere in the annals of the Academy, we find as donors many of the most eminent persons in the land, including the royal family. Within its rather tight confines, the Museum was becoming a national institution.

The interest of the public in oddities and anthropomorphic

displays was also accepted by the curators. Sven Nilsson did admittedly oppose placing the Museum's male lion in a heraldic position, but he was delighted with a group of 14 monkeys from the Grill collection purporting to represent a courtroom scene. Fries comments on one occasion on an offer from the Academy's North African correspondent Johan Hedenborg: "In my opinion such a Museum showpiece as a Giraffe ought, if it is to be worth its place, to repay the sum spent on it, however great this may be. And this it does at once with admission charges to view it." He calculated that 3,600 visitors would be enough for it to break even, but that at least half the population of Stockholm would be inclined to visit.[10] One may wonder what such a giant as the giraffe—visible above the throng in all the pictures of the Museum's exhibition—did for Lamarckian thinking. How could the creature have got a neck like that?

One spectator's reaction is recorded in the diary of Mrs. Märta Helena Reenstierna, lover of natural history and horses, who lived at Årsta near Stockholm and visited the Museum in July 1833, before the giraffe arrived, but after the new premises had been opened:

"On Wednesday at 10 A.M. I and Mamsell went to town. We drank chocolate with my nephew, who accompanied us to the Academie of the Sciences in the Westman House, where we saw shells, birds, fish, land and sea animals, apes, snakes, petrifactions etc. etc. as few could describe and least of all I. The late Prince Fredric Adolph's horse, with its remarkable mane and tail, which I had seen while he was alive, particularly appealed to me, and now I was able to approach him. An Angora billy goat, some reindeer were also nice to see and stroke, but in everything—everything—could be seen the great power of God. The number of the birds, their form and manifold colors surpass all human description. Mamsell admired the tail of the peacock, but it was like those I have had and so did not surprise me. But oh, how great is the Lord who everywhere so shows his omnipotence!"[11]

We notice that the lady of Årsta saw nature in the eighteenth-century manner, as a manual of God's greatness, a physico-theology. She wondered, but the cause of her wonder was the unfamiliar, not the peacock, since she had had one on her estate. Fries, too, saw his museum as a paradise, but it was an increasingly crowded paradise, in need of order and scientific description. Fries was therefore the first to have a guide to the collections printed (1836).

The Reorganization of 1841; Expanded Activities

In 1841 the Museum was given the official name of *Naturhistoriska riksmuseet* (the Swedish Museum of Natural History). It now had five sections, and government support was increased. The Academy's own collections were thus incorporated in those that the state had accumulated. An order of the king-in-council on January, 25 1841 gave a new budget: a curator with overall responsibility at a salary of 1,600 *riksdaler* banco and four section keepers (800 *riksdaler* each), together with taxidermists, two students, one artist, and 2,200 *riksdaler* to enlarge the collections. Altogether 7,500 *riksdaler* and a staff of about a dozen against the previous three or four employees. The estimates of March, 20 1858 were for two senior keepers at 4,000, three junior ones at 3,500, a taxidermist receiving 1,500, students, artists, purchasing, a total of 33,850. Just over a third for purchases, whereas the proportion had been just under a third in 1841. In 1877 the estimates were 41,200 for salaries and 33,420 for purchases, expenses, etc.—this side now being much better provided for. Additional sections were added for palaeontology (1864) and palaeobotany (1884).

Many of the leading names in Swedish natural history passed through these rooms. The reorganization of 1841 tied Carl Jacob Sundevall (1801–1875) more closely to the Museum; he was first and foremost an ornithologist, but also the author of a survey of the animal world of Aristotle that is used to this day (*Die Thierwelt Aristoteles,* 1864) and of an early work on phonetic script. Sundevall was mild-mannered but efficient, and played an important part in the passing of Swedish fishery legislation, a field in which, as in its contact with sporting interests, the Museum could do work of visible benefit to the community. Also in 1841, the marine zoologist Sven Lovén (1809–1895) began his long career as curator of the invertebrate section. At a later date, we find, for example, Nils Johan Andersson (1821–1880), who charted the flora of the Galapagos Islands; the mineralogist Carl Gustaf Mosander (1797–1858), discoverer of the element erbium; and his successor as curator of the mineralogical collections, Adolf Erik Nordenskiöld (1832–1901), renowned as an explorer and a cultural figure, whose section also housed a chemical laboratory. We shall come across other important names.

One noteworthy characteristic of the curators is the young age at which they were appointed. Nordenskiöld was 28 (1858), Fries 32 (1831), the entomologist Carl Stål 34 (1867), his successor Christopher Aurivillius 30 (1883), the palaeobotanist Alfred G. Nathorst 34 (1884), Sven Lovén 32 (1841), the zoologist Fredrik A. Smith 32 (1871), Nils Johan Andersson 35 (1856), Sundevall 38 (1839), and the zoologist Einar Lönnberg 39 (1904). Naturally there were those who were older when they took up their posts, but this was usually because there had been no room for them earlier. Good scientists were in fact taken almost as soon as they had obtained their degrees. There was no application procedure comparable with that normal elsewhere in Swedish academic life, even though the curators enjoyed the title of professor. The curator had to be able to stand up to the rigors of scientific expeditions, as well as the heavy work of organizing the Museum; the job was physically demanding. In its best periods, the Museum could radiate youthful vigor.

The founding of the palaeobotanical section in 1884 was a new departure. Its background can be traced to the polar expeditions of Nordenskiöld and Gerard De Geer and their findings of tertiary plant fossils. These had been preserved in various parts of the Museum. From what we know of the discussion, the personal position at the Museum that Nordenskiöld proposed might be interpreted as a sinecure for Nathorst, who had discovered the fossil flora in the coal formations of Skåne as a mere 20-year-old in the 1870s. An unusual argument was advanced in Nathorst's favor in the debate on the matter in parliament in 1883: "He is, as the gentlemen probably know, deaf. But it seems almost as if, when his ear was closed, his eye was sharpened for this distant world, of which we can still see something but from which nothing can any longer be heard." However, it was Nathorst's competence rather than the members' sympathy that decided the matter in his favor. The fear was voiced that he might receive an offer from abroad and be lost to the country. He had also made the remarkable discovery that Sweden had had a southern flora, that Greenland once really had been green with sycamores, planes, and magnolias. Altogether the Riksdag devoted 50 closely printed pages to the discussion instigated by Nordenskiöld.[12] The post was created against the recommendation of the standing committee of supply and became part of the regular establishment in 1910.

The history of a museum is largely that of the objects it contains. Help in collecting came from all sides, from forest rangers, district medical officers, students, apothecaries, travellers, adventurers. The celebrated writer Fredrika Bremer donated petrifactions from the Levant and North America (1862). In 1904 the envoy Herman Wrangel produced an okapi, a species only recently discovered. Sweden's lack of colonies was to the Museum's disadvantage, but during the period when St. Barthélemy was a Swedish possession, a good deal of Caribbean material arrived (from Samuel Fahlberg, Olof Swartz, Axel Goës, etc.), the examination of which probably merits a study of its own. The contributions of Swedish consuls in different parts of the world were valuable; the facilitating of this type of trade was evidently part of their duties. Material from missionaries was also important. Moreover, the Museum also dealt in specimens itself, rather as Fries had once proposed. Animals that had died in visiting menageries were accepted as donations or purchased. Exchanges were particularly important in shaping the collection: duplicates were sent to Kew, the Smithsonian, or Paris in return for species that were needed. Nils Johan Andersson brought in from Lapland some 40,000 plant items intended as exchange material. To describe it, a *Flora Lapponica exiccata*, was produced in an edition of 15 copies in 1865. Another feature of the Nordic material was the expensive concentration on whales, with purchases from East Finnmark in 1865, a finback whale from Bohuslän in 1880, and on another occasion two stray bottlenose whales that had been killed in the Baltic. In particular the collections were augmented by the results of deep-sea trawls from the marine biological station Kristineberg north of Gothenburg and elsewhere. There were also significant additions by purchases of the private collections of the curators and other members of the staff—such collections existed, of course, although expressly forbidden by the statutes; the prohibition presumably applied to the arrangement of private collections through the Museum and to trading for private profit.

The many expeditions supported by or otherwise involving the Academy were of great importance to the Museum. The Linneans' voyages of exploration with the Swedish East India Company are an early and classic example. It is harder to form a clear picture in the nineteenth and twentieth centuries. Many travellers both in Sweden and abroad applied for grants, many departed, and the majority paid by delivering material. Among them may be mentioned Johan Hedenborg in Africa, the eccentric Claes Mellenborg in Java, Johan Vilhelm Zetterstedt in Lapland, Gustav von Düben around the East Indies, the luckless Wahlberg in South Africa; the round-the-world voyage of the frigate *Eugenie* in 1851–1853 under Philip Virgin and Carl Skogman and with Hjalmar Kinberg and Nils Johan Andersson as its zoologist and botanist respectively; the circumnavigation by the steam frigate *Vanadis* in 1883–1885; all the polar expeditions; Nordenskiöld's transit of the north-east passage with the *Vega* in 1880–1881 and his exploration of Greenland; Carl Skottsberg in Tierra del Fuego, Sven Hedin in Asia, and so on. Material collected by the great British *Challenger* expedition also reached the Museum, whose tentacles reached far down into South America. Regnell's herbarium, donated by the wealthy Anders Fredrik Regnell, who was a doctor in Brazil, and steadily extended, has become an institution within an institution. The travellers brought research material and forged scientific contacts in all parts of the world. Nor should we forget the glamor and sense of adventure that their voyages created for the Museum. They were in every way the best of publicists.

At Berzelius' suggestion, the Museum had been supplying the gymnasia with material for natural history studies since the 1830s. Much of this came from the stock of duplicates, of course, but it also meant that some collections were split up, including parts of the Paykull collection (which the statutes had decreed should be kept together and which only became part of the Museum in 1849 after having previously been kept separate). Other collections used were Westin's from Brazil, Hedenborg's from Egypt, and Wahlberg's from South Africa. This policy of distribution was both an expression of educational idealism and a response to constant overcrowding. Material was sent to schools in Falun, Härnösand, Västerås, Hudiksvall, and Östersund, where the pupils could enjoy the sight of exotic fowl from Brazil, lizards from Africa, and much else.[13] In due course such adornment of the old school museums might give valuable stimulus to young hearts and minds. The schools may therefore with only a slight exaggeration be said to have been both recruiting grounds and branches of the Museum.

The Growth of Research

There seems to be no real answer to the question of when research became a self-evident part of the work of the Museum of Natural History. It is clear that collecting, arranging, and classifying, which may appear relatively humdrum tasks, present continuous and difficult scientific problems. Over the indistinct boundary between collecting and science, research gradually found its way into the work of the Museum. Sven Lovén's studies of molluscs fall into this category and should not be categorized as popular exhibition zoology or as part of another important tradition of the Museum, namely, the interest in hunting, reflected in the concentration on mammals and birds. We have also mentioned a kind of utilitarian research by Sundevall in the field of fisheries. Much naturally depended on the staff concerned—the possibilities were demonstrated in the Musée d'histoire naturelle in Paris in the earlier part of the nineteenth century with Lamarck in charge of the conchylia, Geoffroy St. Hilaire of the mammals, Cuvier of the palaeontological material, Lacépède of the fishes, Le Vaillant of the birds. With such a star-studded cast, there is no questioning the description of the museum as a creative scientific environment.

A corresponding period appears to begin at the Swedish Museum of Natural History around the middle of the nineteenth century. One episode from a process that is otherwise hard to picture clearly, an item of business: in 1880, the young Christopher Aurivillius, who was to become the secretary of the Academy 20 years later, had been nominated to succeed Carl Stål as entomologist. This was eloquently opposed by the lecturer in Lund, C.G. Thomson, at age 60 twice Aurivillius' age and with a work of over 3,500 pages on Scandinavian beetles, written in impeccable Latin, to his credit, whereas Aurivillius' style in his writings had been careless. One reply was from Sven Lovén, who maintained that the duty of the head curator was to look after his section, promote its development, and publish writings consistent with "the primary purpose of the Museum, which is scientific, and involves not merely constant description of species but also biological study in the modern sense." Thomson pointed with some justification to the instructions to curators, which do not contain a word on "any writings of such a kind."[14] He might also have asked what was meant by the expression "in the modern sense." The Museum was quite simply not content to be "museological." At the same time, a more pronounced emphasis on research qualifications was evident in professorial appointments at Swedish universities, although this emphasis was not clearly formulated in the university statutes until 1916. We may also note that it was Lovén who replied on the Museum's behalf, although it was not strictly his place to do so. He, more than anyone, was the Museum's ideologist, his authority enhanced by his status as one of the eight corresponding members of the French Académie des Sciences.

Research was the purpose of the various zoological stations established in the second half of the nineteenth century, a French one at Wimereaux in 1873, a German one in Trieste in 1875, a British one at Plymouth in 1887, and a Norwegian one (by Fritiof Nansen) in Bergen in 1891. This movement may be said to have received its manifesto in A. Dohrn's *Der gegenwärtige Stand der Zoologie und die Gründung zoologischer Stationen* (1872), which warned against overestimating the value of "museum collecting." Back in the 1830s, Fries and Lovén had examined the marine fauna at Kristineberg on Gullmar Fjord on the west coast of Sweden, and with Regnell and Gustaf Retzius providing the finance and Lovén as director, a marine biological station was set up there in 1877 under the auspices of the Museum (cf. C.G. Bernhard's article in this volume). An ecological approach, to some extent in opposition to conventional museum work, thus started to enter the picture—a change that was paralleled by the growth of aquarium-keeping as a hobby. But the role of Kristineberg in the history of Swedish biology is not confined to the purely scientific plane. The pleasant summer months spent among gruff fishermen and sprayswept skerries laid the foundations of lasting brotherhood and friendship between tutors and white-capped students.[15]

The theory of evolution was the great watershed of the period in biological research. In view of the opposition of such bodies as the British Museum, a cool reception might have been anticipated. In London Richard Owen, John Edward Gray, and Albert Guenther were all sceptics for many decades. The question of the existence of the species was a vital one in the museums, whose very work was based on the collection and classification of species and was published in the form of comprehensive catalogues. In such a clearly defined world, Darwinism could appear highly theoretical and impractical. But this was not the case in Stockholm. Darwinism established its first Swedish bridgehead at the Museum of Natural History. The botanist Nils Johan Andersson's travels had taken him to the same places as Darwin, and on the Galapagos Islands, he had been struck by the same realization as Darwin, although in Andersson's case with relation to the flora, the subject of his doctoral thesis. The annual report of 1854 describes his collections on behalf of the Museum, "among which those from the Galapagos Islands are probably for the most part new to science and exceed in number those previously brought thence to Europe." Anderson visited England in 1857, was in touch with Hooker, and became an early protagonist of Darwinism. Only four months after the publication of *Origin of the Species* in 1859, Sven Lovén reported on its contents; in his case the possibility of a realignment of science was combined with political radicalism. Gustaf Lindström, who was in charge of the palaeontological section, translated Darwin's *The Voyage of the Beagle* (1879). C.J. Sundevall in the vertebrates section appears to have accepted Darwinism in 1864, and Carl Stål mentions it with approval in 1869. "The Academy of Sciences remained the bastion of Darwinism in Sweden in the 1870s," it has been said.[16] One of the most important evolutionary biologists in Sweden was F.A. Smitt, active at the Museum from 1871, who also showed (neo-) Lamarckian sympathies—the combination was not uncommon up to the end of the century. Later the Museum's staff was joined by Gustaf Retzius, who demonstrated an equally strong commitment to Darwinism. One can only speculate why the Museum's so rapidly accepted the new theory, but two explanations present themselves: the freedom from a traditional academic milieu characterized and fettered by idealistic philosophies—such as prevailed in Uppsala—and the relative youth of the staff. Taxonomy and the theory of evolution were in fact easy to combine, as the Swedish example shows.

A greater threat to the Museum, in the long run, came from laboratory biology, represented in Sweden by Vilhelm Leche, who established a department of zootomy at Stockholm Högskola just before the end of the century. His constant need of fresh animals sometimes met with incomprehension from the naturalists of the old school. "The Museum of Natural History is full of fresh animals—use them," they would say. "You can't dissect stuffing," retorted Leche. The Museum did not cater for the most interesting aspect—life itself. Alluding to the Latin name for cockroach, he bestowed on the advocates of descriptive science the unflattering appellation of "bobbologists."[17] In its darker periods, the Museum has, like other similar institutions, lapsed into the state of a mausoleum of a mummified nature, tended by a grizzled priesthood, and pervaded by an odor of formalin from the past.

Portraits of the Museum from the inside are scarce. The memoirs of Gustaf Retzius, which unfortunately are sternly edited, contain long lists of members of staff, all referred to in equally complimentary terms. Retzius was indeed the friend of everybody, but he reports that there was at first some tension between Lovén and Nordenskiöld. The exiled young Finnish professor was seen as rather haughty and inconsiderate. Retzius brings out his truly splendid qualities, that he really worked not for himself "but for higher, idealistic ends, *usually* [sic!] for scientific goals." Some of his ideas were, of course, somewhat fanciful, but as he himself maintained: "A *little* humbug does no harm if you have a goal to reach." The following episode conveys the mixture of reserve and intimacy in the Museum before the move out to Frescati. The curators worked fixed hours in the building where Nordenskiöld and the secretary of the Academy lived. On one occasion Nordenskiöld was to arrange dinner for the staff of the Academy and the Museum and wrote out the invitation cards himself. The caretakers waited, dressed as servants, but the only people who arrived were the secretary, Georg Lindhagen, and Professor Erik Edlund, who had been invited by word of mouth, since they lived in the building. The invitation cards were then discovered in the host's office, undispatched.[18] Nordenskiöld was in other words an archetypal professor, both grand and absent-minded. As a national hero and member of the second chamber of the Riksdag, he could lend luster to the Museum and further its cause. He was a great Swedish patriot, even though with Finland in his heart.

The Meeting with the Public

We have already had a glimpse of the Museum of Natural History as an institution open to the general public back in Sparrman's time. The first truly public museum in Sweden and one of the first in the world was the Royal Museum in the north-east wing of the palace, which was opened to the public in 1794 and originally concentrated mainly on classical art. The parliament of 1844–1845 granted appropriations for a separate building, which led eventually to the opening in 1863 of the building now known as the National Museum of Fine Arts.[19] Discussion of the future site of the Museum of Natural History took place at the same time, but with little immediate result.

The taste of the age was for exhibitions of all kinds. A museum of natural history had been founded in Malmö in 1841 and another, incorporating earlier collections, in Gothenburg in 1861, both as sections of larger museums. In Stockholm a kind of zoological garden had been planned in Humlegården in the 1830s; later, in 1880, an attempt was made to obtain support and finance for a more substantial establishment on Djurgården, adjacent to the Nordic Museum. The efforts were unrewarded on this occasion, but Artur Hazelius was more successful with his Nordic open-air museum Skansen—which included wild and, initially, exotic animals. Hazelius' Scandinavian ethnographic exhibition could be viewed from the 1870s onwards, and may be said to have blazed a trail for the ethnographic exhibitions of the Natural History Museum. Sweden also had its versions of the great world exhibitions, which were apt to include both human and animal oddities and rarities. At the Stockholm Exhibition of 1866, the "Malm whale" was on display; this was a young blue whale whose owner, the whimsical curator of the Gothenburg Museum, August Wilhelm Malm, believed it to belong to a new species, which he named *Balineoptera Carolinea* after his wife, "the person who is to me the dearest of all." The visitor could pass through the jaws of the whale like a second Jonah, take a seat in its stomach, admire the starry blue and yellow roof, and enjoy a cup of coffee. In Stockholm public interest was high, and before the exhibition closed 37,000 visitors had paid to see it, figures with which the Museum of Natural History could not compete.[20]

A competitor in the same field, the Biological Museum, was completed in 1893; this was a building in traditional Nordic style presenting a series of dioramas in various Nordic biotopes. The number of species was high, thanks to the expertise of the taxidermist Gustaf Kolthoff, although not quite the 4,000 that was claimed, and the displays were enlivened by the backgrounds

The main hall of the Museum of Natural History in the Westman House, ca. 1900. (Courtesy of Stockholms Stadsmuseum.)

painted by Bruno Liljefors and Gustaf Fjaestad. The Biological Museum, nowadays somewhat dilapidated, although still a remarkable monument to the new museum philosophy, won immediate acclaim and attracted large crowds during the Stockholm Exhibition of 1897. It bore witness to the contemporary interest in ecological aspects; the plants and animals were not merely to be placed in a row, they were to be brought to life by the surrounding environment.[21]

Museums of natural history can obviously be arranged according to various principles. The oldest is according to *rarity:* we want to see whatever is biggest or first, rarest or most beautiful. But an exhibition might also have *utility* as its theme. In an otherwise little-known Museum of Science, Crafts, and Art, the Stockholmer of the mid-nineteenth century was presented with a comparative display of natural objects province by province (including in the Swedish area the landscape of St. Barthélemy), and the uses to which they could be put.[22] The ideal of instructive lucidity was seldom so clearly expressed, but the Museum of Natural History, too, found room in its botanical displays for economic utility, including material transferred from the Stockholm Exhibition of 1866; otherwise the plant world was hardly exhibited before the new order of 1916. Or, of course, the arrangement might be *taxonomic,* continually vitiated by the fact that the organisms, like books in a library, vary in size. During the nineteenth century, the museums generally followed the taxonomic line—although naturally not rigidly. The ecological approach was as yet difficult to realize, as was the historical dimension, so evident in the prehistoric lizards of the British Museum—all that was offered in Stockholm were a few bones and plastercasts, some reproductions, and fossil plant material. Nor was much use made of the historical impact and the possibility of identification that could have been offered to the viewer by the species *Homo sapiens,* although the ethnological collections were in the building or stowed away in some nearby storehouse. As a collection for viewing, the Museum had persistent problems as long as it lacked elbow room.

A press reaction of 1904 is illuminating: it was reported that Skansen was the most popular of Stockholm's museums, followed by the Natural Museum of Fine Arts and the Museum of Natural History. The latter particularly appealed to younger visitors. In the entrance hall, one was confronted by the Greenland meteor, "which after its vagrancy in outer space has found the rest that we can offer it—until one day in the fullness of time it will be the turn of our planet to be shattered to smithereens and restlessly to wander the universe." Then the writer went through room after room, inspecting wooden models of whales—the critical visitor might feel that such painted wooden substitutes were evidence of a certain paucity in a national museum of a country that almost borders the Arctic Ocean. And perhaps he wondered whether the rest of the animal collection also consisted of "wooden blocks." Critical comments on the motheaten state of everything were interspersed with an appreciation of the diversity. One was struck by the big glass case containing the group brought home from Greenland by Nathorst, a family of musk oxen—father, mother, and baby. Our literary-minded guide inspected the Museum's "rogues' gallery," i.e., the most unpleasant-looking creatures, the snakes and crocodiles. "A new room and we descend into the depths, . . . where fish swim around in eternal silence, their eyes perpetually locked open as if in constant terror." Oh well, here in the cases were no predatory fish behind every stone, it was very peaceful in the cases. "Here the swordfish bears his sword in vain." Now to the denizens of the air: "There are the ostriches, towards which every lady visitor directs her steps." The birds retained their color and form, "but where is the movement, where is the happy warble of this avian flock?" Suggestively and mockingly, the author of the article pointed out the Museum's possibilities and deficiencies and concluded with one of the cardinal questions of museology:

> "The whole of this collection, which is now properly locked in the room's many glass cases, has once warbled, buzzed, twittered, grunted, cackled, bleated, whinnied, snuffled, crowed, quacked, hissed, growled, barked, mewed, mooed—I am a little prolix for the purpose of showing that animals, too, have their Tower of Babel. The evil spirit of death has now laid his hand on them and they can move no more. And at last one becomes so used to being in the world of the lifeless that when one comes out onto the street again and sees a sparrow, one is almost waiting to see him motionless with a label tied to his feet. But he definitely does not covet this distinction as he hops carefree about on the cobbles."[23]

When the Museum opened in 1831, it was visited by 100 persons per day. During April–October 1835, only 1,185 visitors were recorded, and for 1836 the figure was just under 2,000. The number of visitors in 1840 was still 2,000, including 56 with the complimentary tickets issued to schools and students. In 1848 the figure fell as low as 1,400, but in 1857 the admission charge was reduced to 25 öre with free entry on Saturdays, and the number of visitors rose. The difference that this made, together with, perhaps, an indication of the public's social background, is apparent in the fact that in 1861 there were 1,111 paying visitors on Wednesdays, whereas on the free day the figure was ten times as high. The Museum was closed for repairs between 1861–1866. After 1866 the figures look different, with new opening times (including Thursday and Sunday) and between 1,000 and 2,000 on Sundays according to the annual report. After a few years, however, figures dropped again, falling as low as 455 paying visitors; the number of nonpayers may certainly have been many thousand. The longer hours of opening also enabled working people to visit the Museum, but the extent to which this took place is unknown. The figures fell significantly during the winter—the cold climate of Scandinavia has its unfavorable effect on the study of nature. The underlying educational ideal was sometimes heard in the ever enthusiastic annual reports: "From which it appears that the public demand for a knowledge—often only superficial, yet always useful and ennobling—of the wonders of nature is not mere passing fancy but a phenomenon of a deeper and more lasting character" (1879).

From 1890 onwards increasing numbers of schoolchildren and soldiers were piloted round the rooms. Rather strangely, the instructions stipulated that the caretaker, assisted by the taxidermist, was to be in charge of these tours. In 1904 the records show that 24,600 saw the zoological collections—which were the ones most often concerned, although other sections, too, might often have tours that were not recorded. In 1915 the Museum was closed for removal. When the new Museum opened in 1916, the attendances rose, as was expected. The figure in 1917 was 72,000, larger than for any Swedish museum except the National Museum of Fine Arts; as a hybrid of park and museum, Skansen need not be included. But it could have been higher, and the drive to attract the public could have been more energetic. The head curator,

Einar Lönnberg, complained of the poor transport facilities, which kept people away from Frescati. A comparison with the British Museum and its new premises in one of London's green spaces may be unfair; the latter had approximately 400,000 visitors per year during the 1880s and 1890s.

New premises and new attractions led to increased attendances. In 1907 a "particularly fine donation" was received from Anna Broms, widow of one of the great patrons of the Museum, the tough industrialist Gustaf Emil Broms. The four stuffed gorillas, of which the large male was an especially beautiful specimen (of the rare sub-species *Gorilla gorilla matschei*), boosted the figures; the gorilla family was to remain one of the Museum's main attractions for decades, together with the musk-ox family from Greenland. It is tempting, bearing in mind the prevailing family ideal, to speculate on the anthropomorphic comparisons that such displays may have provoked. Of course, feelings of all kinds were aroused, and there was an aesthetic experience, too. The Museum was also used for animal study by such artists as Georg von Rosen (1906) and Carl Milles (1910).

To oblige the visitors, small guidebooks were produced. A window on the world was opened in 1906 with the start of the journal *Fauna och Flora,* under the editorship of Einar Lönnberg. The journal formed and has remained an important bond between the Museum and a large public. It has given pride of place to the Swedish fauna; despite the title the flora has been glimpsed only more fleetingly, perhaps because this has been catered for by the older *Botaniska notiser,* perhaps also because the Museum's prime concern has been zoology. The part played by *Fauna och Flora* in Swedish popular education is worthy of a study of its own. It may be regarded as a counterpart to Sven Nilsson's old *Tidskrift för jägare och naturforskare,* (mentioned above), with a characteristic contemporary emphasis on the Swedish and the animal protection aspects.

The New Museum of Natural History

The accommodation problems of the Academy were largely synonymous with the ever-growing needs of the Museum. Initially the Lefebur House on Lilla Nygatan was used, but with the Paykull donation new arrangements became necessary. In his deed of foundation in 1820, the king had suggested Manilla on Djurgården, to which the Academy had objected on the grounds that it was too far from the center. In 1828 the collections were moved to the Westman House near Adolf Fredrik's Church. A series of expansions took place in the same block until further extensions were refused by the Riksdag. The overcrowding, together with the disturbing lights and noise of the site in the very center of Stockholm led in 1900 to the appointment of a building committee. Sights were first set on Djurgården, but unlike his predecessor Carl XIV Johan, Oscar II was against building there—although the persistence of Hazelius led to permission being granted for the Nordic Museum to move there in 1880. From 1904 onwards the Frescati district north of the city was the only feasible choice. It had the advantage of proximity to the Bergius Garden, to the testing grounds of the Royal Academy of Agriculture and Forestry, to the Geological Survey of Sweden, and to planned Nobel institutes, all of which formed part of the new "science city." There, too, the Museum of Natural History was to grow up, after the many twists and turns of a story that can here be told only briefly.[24]

For the grandiose plans of the committee, land was needed, and some members would have gone even further and, for example, provided space for curators' residences. But the proposal was clearly at odds with the idea of the Museum as a museum of the people, easily accessible in the center of the capital. Instead, the conservation and research aspects were given priority. A century of overcrowding and the awareness that there is really no limit to the creations of nature, as intimated above, had taught inescapable lessons. Mention was made of the new Museum für Naturkunde in Berlin, whose director had already asked for new premises "in a place where this building can be expanded *for centuries.*" And from the curators' side, with the solitary exception of ethnography, the interest in research was stressed in a memorandum as early as the 1890s. It was admittedly proposed that the Museum should be open to visitors throughout the hours of daylight, that "exquisite specimens" should show in an easily grasped manner their external and internal form, evolution (!), way of life, geographical distribution, palaeontological occurrence, all arranged in an educational manner to enable the viewer to "derive the degree of basic knowledge of nature that contemporary culture demands." But for scientific work, on the other hand, the collections were to be arranged systematically. And "the greater area is therefore the due of this department."[25] Outwardly the new Museum was already a powerful demonstration of the expansive force of Swedish science at the turn of the century. But unlike its counterparts in London, New York, Berlin, Paris, and Prague, it was almost out in the country. Even today Stockholm has not entirely surrounded the Museum of Natural History.

The scope bears witness to ambition and capacity, to power and money. The Museum was steeped in nationalistic history, and at the great Linnean bicentenary in 1907 (and the Swedenborg Jubilee in 1910), the curators showed their firm adherence to historical tradition.[26] Nevertheless, the old ideal of embracing the whole world of natural history was clearly losing its hold. The Enlightenment dream of the Museum as one vast encyclopaedia was proving ever more difficult to realize. The argument among the curators over the location of the different sections was long and heated, with two opposing camps: the zoologists under Lönnberg, the head of vertebrates, preferred unity; the botanists under Nathorst recommended individual pavilions. The outcome may be called a compromise, with a main complex and a separate building for botany. Everyone spoke enthusiastically of the need for expansion and for links between the units, but in practice the same rhetoric could clearly lead to different conclusions. The story of life as explained by the laws of evolution could be studied in an integrated museum of natural history, but, on the other hand, the life sciences had gone through an inexorable process of specialization, which made the "pavilions" logical.

In another way, too, the comb-shaped pattern that the buildings described bore witness to the new day that was coming—but had not yet fully arrived: electric lighting was, of course, a part of it (the Museum also had five electric lifts from the day it was built), but there was not enough for every need. The architect had had to put the buildings a good way apart to admit as much daylight as possible to the high-ceilinged rooms. The whale museum, for example, had a ceiling height of 7.6 meters and windows accordingly. It was for light, among other things, that the curators fought like coral on a reef. The size of the Museum is thus a result both of the technical knowhow and shortcomings of the day.

One expression of the divergent tendencies was the reluctance of the ethnological section to move out of the old building. It had long fought for its independence. In 1871 the Academy made the first move towards founding an ethnographical museum.

The new building of the Swedish Museum of Natural History, opened in 1916.

For the time being, the collections had been placed under the vertebrate section, and from 1880 they had been displayed separately twice a week at a charge of 25 öre. Interest in ethnography was growing: Hazelius' Nordic collection, the Swedish Society of Anthropology and Geography (1873), the much-publicized orientalist congress of 1889, people of exotic race visiting as popular entertainers, etc.; in 1907 a major ethnographical exhibition was held, most of the exhibits having been collected by Swedish missionaries.[27] The collections were undeniably remarkable and worthy of proper premises. The curator, Hjalmar Stolpe, the Swedish pioneer in the subject, asserted firmly that the Museum had the duty to be accessible to the public and to educate. The section simply refused to go—which led to a long period of semi-obscurity, only partially relieved by its being freed to become an independent museum in 1935, still without proper premises. Yet on the main plan of the Museum of Natural History, the section is still there—shown, indeed, as a large wing. Here the idea of the Museum in its totality lives on; here, too, can be seen the position ethnography had reached in Sweden at the turn of the century.

The architect was Axel Anderberg, well-known from other similar, if more modest projects. The building, which was erected between 1906 and 1915, cost approximately 3.3 million kronor, to which was to be added 600,000 for interior decoration, largely financed by the sale of the previous property. The result was not only handsome, but also pure in style, with traces of American campus architecture. The element of nationalism was relatively unobtrusive, but it was nevertheless inevitable as a kind of "superideology" because of the concentration on Swedish nature and Swedish science. We may note the avoidance of nationalistic symbolism—the money had to be saved for other things; one concession was a pair of sculpted bears and eagle owls at the entrance. "The greatest simplicity" was the theme— walls and ceilings, with a few exceptions, were smoothly plastered, and the tower building in the center was actually intended as a reservoir, although it was also decorative. The terraces were of granite, while the buildings had brick and granite facades of grey and red. At this time Stockholm was rapidly expanding with brick as the predominant material—the Royal Institute of Technology represents the same period and style, as does the stadium built for the Stockholm Olympics of 1912. The

Museum of Natural History, one of the largest public buildings to have been built in Sweden, gave the Academy cause for pride. When they moved in, the officers of the Museum must have faced the future with great confidence.

The Twentieth Century

The history of the Museum after 1916 will not be discussed here. A new era had begun, with new premises and a higher priority being given to research. The Museum has fanned out, in material possessed, in finance, and in the size of its staff. The budget in 1914 amounted to 215,000 kronor for a staff of 36, in addition to which was a large nonpermanent workforce, to be compared with one of only two or three individuals when the Museum opened. Corresponding figures for the period since World War II are 1 million kronor for the budget of 1948, 1.6 million in 1956; the number of employees about 60 in 1948, 66 in 1956. The number of visitors has been fairly constant: 50,000 in 1934, 68,000 in 1948, 62,000 in 1966. The largest number of visits take place in late spring, the time for school trips. It may be added that it was in particular the whale museum—the biggest in the world when it opened before the other sections in 1908—that demonstrated literally the vastness of nature to generations of Swedish schoolchildren. But there is no denying that its position out in Frescati has not helped the Museum's attendance figures. Closer examination would probably show that the demands of popular education and research have continued to conflict. But research has flourished. According to one estimate, the staff of the Museum produced during the 20 years from 1906–1925 807 works filling 27,564 pages, figures that also testify to a growing administrative burden.[28] Since 1965 the Museum has continued to exist as an institution independent of the Academy, albeit with some difficulty. If Alfred Nobel had decided to award a Nobel prize for biology, this might well have provided constant inspiration.

The ties with the parent institution were also more literally severed at about the same time by the building of the motorway from Stockholm north-eastwards towards Finland and the U.S.S.R. The huge campus created early in the century has now been divided; Axel Anderberg's stone temple has been isolated. In its present form, the Museum continues to be a majestic and indispensable center for maintaining the Swedes' knowledge of nature and for international research into the diversity of creation. The handy cabinet of natural history has become a palace. The Museum of Natural History on the Frescati plain is, of course, a museum of the diversity of nature, but also a museum of itself, of older collections, of the development of museology, of the work of its staff, and last but not least of the meeting between man and nature in the modern age.

Notes

1. There is no overall history of the subject. But compare parts of Ilse Jahn, Rolf Löther, et al., *Geschichte der Biologie, Theorien, Methoden, Institutionen, Kurzbiographien* (Jena, 1982). See also the special number of the *Journal of the Society for the Bibliography of Natural History* 9:4 (1980).
2. Cf. Gunnar Broberg in the introduction to *Rudbeck's Fogelbok* (Stockholm, 1986).
3. Charles Coleman Sanders, *Mr. Peale's Museum* (New York, 1980).
4. Cf. Lynn Barber, *The Heyday of Natural History 1870–1920* (London, 1980), 152–168. William T. Stearn, *The Natural History Museum at South Kensington* (London, 1981); Sally Gregory Kohlstedt, "Curiosities and Cabinets: Natural History Museums and Education in the Antebellum Campus," *Isis,* 79 (Sept. 1988.), 405–426.
5. Cf. E. Stresemann, *Die Entwicklung der Ornithologie von Aristoteles bis zur Gegenwart* (Aachen, 1951), 271; also G. Broberg, "Natural History and Encyclopedism" (forthcoming). The number of insect species was stated by E.O. Wilson in *Science,* December, 13 1985.
6. Cf. Lindroth, *KVA Historia* 1:1, 620–629. For the earlier period, see also Yngve Löwegren, *Naturaliekabinetten i Sverige under 1700-talet* (Lund, 1952). The major work on the nineteenth century is Einar Lönnberg, et al., *Naturhistoriska riksmuseets historia* (Stockholm, 1916), which is, however, difficult to grasp on account of its wealth of detail and not always reliable. There is no work dealing with the twentieth century. Use has been made of the Academy's detailed annual reports, memoirs, various minor publications, etc., although not of the voluminous correspondence of Lovén, Sundevall, Lönnberg, and others.
7. F. Weber and D.H.M. Mohr, *Naturhistorische Reise durch einen Theil Schwedens* (Göttingen, 1804), 180.
8. Cf. in particular, Yngve Löwegren, "Sven Nilsson zoologen," in *Sven Nilsson: En lärd i 1800-talets Lund,* ed. G. Regnéll (Lund, 1983), on the scientific importance of hunters, 112, 123.

9. Ida Sager, *Bengt Fredric Fries* (Stockholm, 1916), 135 f., 142, 155, 162, 172, 187f, 164, 247 ff.
10. Löwegren, 110 f; Sager, 260.
11. *Årstafruns dagbok* 3 (Stockholm, 1953), 251 f.
12. *Första och andra kammarens Protokoll, 1883, 1884,* nos. 23 and 31. Cf. A.G. Nathorst, "Riksmuseets paleobotaniska afdelning," in Lönnberg, *Naturhistoriska riksmuseets historia,* 245 ff.
13. Information from Yngve Löwegren, *Naturaliesamlingar och naturhistorisk undervisning vid läroverken* (Stockholm, 1974), 34–39.
14. C.G. Thomson, *Svar till S. Lovén rörande Intendentsbefattningen vid Riksmuseum* (Lund, 1883), 11 f.
15. Cf. Hjalmar Théel, *Om utvecklingen af Sveriges zoologiska hafstation Kristineberg* (Uppsala, 1907).
16. Ulf Danielsson, "Darwinismens inträngande i Sverige," 1, *Lychnos* 1963–1964, 201.
17. Mia Leche-Löfgren, *Våra föräldrars värld* (Stockholm, 1934), 83.
18. Gustaf Retzius, *Biografiska anteckningar och minnen* 2 (Uppsala, 1948), 175–192. On Nordenskiöld, see *Ymer,* 1980.
19. Cf. Lars O. Lagerqvist, "Museernas framväxt," *Museiperspektiv;* 25 articles on the museums and society (Rapport från kulturrådet 1986:2), 180–200.
20. L.A.Jägerskiöld, *Upplevt och uppnått* (Stockholm, 1943), 311–323; *Ny illustrerad tidning* 1866.
21. Cf. Allan Ellenius, *Bruno Liljefors* (Uppsala, 1981).
22. P. Hamberg, *Vägledning vid beseendet af Museum för naturvetenskap, slöjd och konst* (Stockholm, 1863).
23. Celestin (= Beyron Carlsson), "Från vetenskapens högkvarter," *Varia, illustrerad månadsskrift* 7 (1904), 193–201.
24. On the building question, see *Handlingar ang. K. Vetenskapsakademiens Nobelinstituts Byggnadsfråga* I–III, *Nya Handlingar* I (Stockholm, 1902–1906); cf. also Retzius, *Biografiska anteckningar* and Carl Forsstrand, *Vid sjuttio år* (Stockholm, 1924).
25. *Handlingar* I, cit. 61, 25 f. *Nya Handlingar* (1906), 47.
26. *Carl von Linné som läkare och naturforskare* (Stockholm, 1907); also German translation (contributions by Lönnberg, Nathorst, Lindman, Sjögren, Aurivillius); *Transactions of the International Swedenborg Congress* (London, 1910).
27. *Etnografiska bidrag af svenska missionärer i Afrika* (Stockholm, 1907); E. Nordenskiöld in *Ymer,* 1906–1907.
28. Tore G. Halle, Head Curator, in a radio talk published in 1935.

TORE FRÄNGSMYR

Swedish Polar Exploration

MODERN polar exploration came to be one of the most spectacular missions of the Academy of Sciences in the second half of the nineteenth century. All the major Swedish expeditions to Spitzbergen and the Arctic were undertaken within the framework of the Academy, even if the Academy did not bear the financial burden of them by itself. Many expeditions were the result of private initiatives, the costs being defrayed from donations and other external sources, but the Academy was the scientific guarantor. The Academy may thus be described as the sponsor of Swedish polar research.[1]

The earliest polar exploration was largely marked by rivalry among the larger nations, and behind this rivalry there were naturally economic interests. Gradually a struggle for national prestige also developed. As early as the 1590s, the Dutch were trying to find the North-East Passage, launching expeditions under Willem Barents that penetrated as far as the north-east coast of Novaya Zemlya. Between 1725–1741 the Russians financed attempts led by the Dane Vitus Bering to explore the sea routes between the northern Arctic and the Pacific and thereby reach America. Beginning in 1818 the British sent a series of expeditions to Greenland, Spitzbergen, and northern Canada in the hope of finding a North-West Passage. In 1821 the British parliament offered a reward to anyone who could reach the North Pole. Whoever succeeded would find a new sea route between the Atlantic and the Pacific and also answer the question of whether there existed an ice-free center around the pole. John Ross, Edward Parry, James Clark Ross, and John Franklin are well-known names associated with these expeditions. Some of them enjoyed a measure of success; others failed completely. The struggle against the cold and the pack ice was magnificent in terms of human endeavor, but was in practice an almost impossible task. Some expeditions managed to winter over; others were forced to return after getting halfway. The trials were monumental. The tragic culmination was reached with Sir John Franklin's expedition in 1845, which simply disappeared in the great darkness. By the time Leopold McClintock's detailed report made it clear ten years later that Sir John and all his men had perished, more than 40 rescue expeditions had been mounted by sea and land—rescue expeditions that, it should be mentioned, produced more new knowledge of the Arctic than any other project.[2]

Commercial and national interests were inseparably combined. If expectation of financial reward was one side of the coin, the nation's appetite for heroes was the other. But a third ambition also began to make itself felt, namely, the scientific. The exploration of the Arctic is an important chapter in the history of geography. It was not only a sea route that was sought, but also the answer to scientific questions. The archipelagoes of Greenland and northern Canada were largely unexplored. The North Pole and the surrounding area exerted a particular attraction. It was now known that a North-West Passage existed, with many different channels, although no one had succeeded in navigating it. Similarly, the North-East Passage was still waiting to be discovered—no one had got further than Barents.

Rapid advances were now being made in all the sciences. For Arctic exploration, developments in geology were particularly important. The new glacial theory provided knowledge not only of the polar regions themselves, but also of how the landscape had been formed. For Sweden, this was of enormous significance. Polar exploration might answer many important questions: on eskers and striae, prehistoric plants and animals, land forms and soils. To Adolf Erik Nordenskiöld, the discoverer of the North-East Passage, geology was the gateway to polar research.

The Swedish polar expeditions therefore, unlike those to which we have previously referred, were undertaken for purely scientific motives. Economic interests played no part, and an element of hero-worship does not really appear until the triumphant voyage of Nordenskiöld in 1878–1880.[3]

The Geological Background

The nineteenth century was a period of geological breakthrough. That the breakthrough came so late was due to a number of causes. One was the inherent difficulty of research: the history of the earth had to be reconstructed, but many of the "sources" for this historical research had long ago been destroyed. Another was that strong religious interests were involved in this uninvestigated field. The Bible contained one version of how the earth had been created and developed, and this could not simply be disregarded when it did not agree with empirical findings.

The period before 1800 is a kind of "prehistory" of the science of geology. The term "geology" only began to gain general currency towards the end of the eighteenth century. Many questions were difficult to solve, especially if they impinged on biblical problems. One such question was of the age of the earth, which the Bible declared to be no more than 6,000 years, whereas Buffon and other natural philosophers were thinking in terms of much longer periods. Another controversial question was the nature of fossils— whether they were the remains of organisms or had been washed ashore by the Flood. A third question was whether the form of the earth had been created once and for all or whether it was constantly changing as a result of processes of erosion and accumulation.[4]

One specifically Swedish question was the problem of the "diminution of the waters." Urban Hiärne had noticed in the 1690s that the relative level of land and water in the Baltic had changed. Emanuel Swedenborg became interested in the problem and was inclined to believe in an astronomical explanation; Anders Celsius took measurements and determined the annual rate of change in the level; and Carl Linnaeus put forward a theory of a biological nature. All seemed to agree that the water had dropped, but the explanations offered were many and various. In the 1760s the Finnish surveyor E.O. Runeberg suggested that the land had risen, but his view was to go unheeded for a long time.

By the start of the nineteenth century, however, all serious geologists were agreed that the Bible and empirical science must be kept apart. This did not mean that all the problems were now solved, far from it. Even if it were accepted that the Bible was not a scientific textbook, many questions remained. Which had played the bigger part in geological events, water or volcanoes? Was the earth's present geology the result of huge catastrophes or slow processes? The historians' portrayal of these discussions has long been greatly oversimplified. The "neptunism" of Abraham Gottlob Werner has been contrasted with the "vulcanism" of James Hutton, the "catastrophic theory" of George Cuvier with the "uniformitarianism" of Charles Lyell. In fact the opinions held cannot be so clearly categorized: it was not a case of two camps struggling to assert directly opposed views. Instead we should think of theories and part-theories, with most geologists trying to form a coherent total view, often by combining elements of different theories. But, of course, there were tensions between different approaches, between different explanatory models. Lyell exacerbated the antagonism by depicting others' theories as antiquated and his own as innovative. He stressed that geological forces described a uniform process, that all the changes took place very slowly, and that the history of the earth was much longer than anyone had yet thought.[5]

Most geologists at this time, about 1830, probably shared Lyell's opinion in its essentials—but not in all the details. It was a question of emphasis. Lyell's critics thought that he laid *too* much weight on uniformity. Against his theory it could be objected that there was room for considerable fluctuations within a uniform process, e.g., through the effect of volcanoes and earthquakes. A new topic of contention arose in the 1830s with the theory that the earth had at some time been through an ice age. Was the ice age in that case a catastrophe, or could it have taken place within a uniform process? The ice age theories were thus among the great scientific issues of the nineteenth century.

A number of geological problems seemed impossible to solve; they concerned the uplift of the land, the formation of the eskers, the spread of erratic boulders, and the origin of all the striae that could be seen on the rounded rocks known as *roches moutonnées.* Lyell assumed that a polar sea had covered much of Europe and that the erratic boulders had been carried around frozen in drifting icebergs. As the climate grew warmer, the icebergs had melted, and the boulders had been deposited on the sea bed. With the land uplift, they had gradually arrived on dry land. The Swede Nils Gustaf Sefström studied the marks on the *roches moutonnées,* and in 1836 published in the Proceedings of the Academy an account of 400 observations. He believed that a huge inundation of stones, a "petridiluvian" flood as he called it, had surged over the land striating these rocks and polishing irregular stones to cobbles. Sefström found supporters in Sweden, including Jacob Berzelius, but no one abroad accepted his ideas.[6]

Louis Agassiz was the first to form a coherent glacial theory from all the previous observations and vague conclusions. He assumed that most of the Northern Hemisphere had been covered by an enormous continental ice sheet that had shattered rocks and put an end to all life wherever it had spread. After first presenting his theory in a lecture at Neuchâtel in 1837, he later developed it in a book, *Etudes sur les glaciers* (1840). Agassiz had studied the glaciers of the Alps and the traces they had left on the landscape, i.e., striae of various kinds, erratic boulders, and moraines. Then he soon discovered that the same kind of traces could be found in other landscapes, far from any glaciers, and that the composition of rocks and soils bore witness to lost connections. Purely empirically such analogical reasoning was not hard to follow. The difficulties were methodological.

One strong objection to the glacial theory was that it assumed that the earth had progressed from a colder to a warmer climate.[7] Against such a development were theories and measurements that appeared to show that the change had been in the opposite direction; the earth had initially been a glowing ball and had cooled down little by little. Another objection was that this was a catastrophe theory, comparable to the legend of the Flood. The theory was considered far too fantastic, particularly at a time when geologists were at last getting away from idle speculation and beginning, almost literally, to feel solid ground under their feet. The most common reaction appears to have been that some parts of the glacial theory were plausible, but that it was impossible to accept all its implications. That there had been glaciers in the Alps and in Scotland seemed reasonable, but the thought of an ice sheet covering the whole Northern Hemisphere was another matter. The very idea stuck in the throat: it contained an element of speculation that was too much for a sound empiricist to swallow.

The First Expeditions

The geological background has been discussed in some detail because the first Swedish polar expeditions were so closely associated with Agassiz' glacial theory that they can only be fully understood in that context. This background also underlines my thesis that Swedish polar exploration—unlike that of other countries—originally had purely scientific motives.

The pioneer of polar exploration was surely Sven Lovén, who was the professor and curator of the Museum of Natural History from 1841. He had become a member of the Academy in 1840, and remained one until his death in 1895. While travelling to Norway, he was given the opportunity to take part in an expedition to Spitzbergen in 1837. The principal object of his studies there was the Arctic fauna, but he also observed the glaciers. On his return home, he examined fossil snails and mussels in Sweden and found that many belonged to species that now existed only in the most northerly parts of the Arctic. With this he tried to prove that Sweden had been through a period of intense cold, a form of ice age. He found support in the new glacial theory, and contributed to its development. He reported his interpretations and conclusions regularly to the members of the Academy, giving lectures and writing articles in its publications. Scandinavia and Finland had, he said, lain under an ice sheet up to a thousand feet thick. When the ice eventually melted, an icy sea formed over Sweden.

As well as doing valuable work himself, Lovén also inspired a younger generation and gave active support in the Academy to polar exploration. The person who was to play the most important part in the application of the glacial theory in Scandinavia was Otto Torell, a pupil of Lovén. While still very young, Torell had been fascinated by Lovén's idea of an Arctic fauna in Sweden; at age 20 he discovered on the west coast of Sweden *Yoldia arctica,* a little mussel that now lived in the polar regions, but was found in Sweden only in fossil form. Torell pondered over his finding. If Arctic creatures such as this had been able to live in Sweden, the climate and natural surroundings must have been Arctic in character; in that case there must have been glaciers in Sweden.[8]

To answer these questions, the young student had to go off and study glaciers for himself. He went first to Switzerland, where he could examine the effects of the ice at close quarters. It was at once clear to him that the glaciers left the same traces as could be seen on the Swedish landscape. But as we have already seen, not all geologists shared his opinion, and this worried Torell: "I myself felt the weight of all the objections to a Scandinavian continental ice sheet so heavily that it took two years of Arctic and Alpine travel to banish the last doubts."[9]

Torell therefore continued his research, paying for his travels himself. He had inherited a small fortune from his father, and financial considerations were not allowed to come between him and his eager research. After visiting Switzerland he set off for Iceland. For six months he and his colleagues undertook a succes-

S. Lovén.

O. M. Torell.

A. Nordenskiöld.

Three generations of polar explorers as depicted in contemporary drawings. From Claes Lundin, *Nya Stockholm.*

sion of arduous and hazardous crossings of ice-covered wastes. In the following year, 1858, he started Sweden's polar voyages with a properly organized expedition to Spitzbergen. Torell's companions were the zoologist August Quennerstedt and the young Adolf Erik Nordenskiöld, who here had his first taste of polar exploration. Throughout the summer they examined the west coast of Spitzbergen, studying plant and animal life, glaciers, and moraine deposits. In the spring of 1859, at the University of Lund, Torell defended a thesis on the molluscs of Spitzbergen, including "a general survey of the natural conditions and former extent of the Arctic region." This work of 154 well-filled pages represents the first attempt to take an overall view of the glaciation of Scandinavia.[10] Previously the main foci of interest had been the Alps and Scotland. Torell showed not only that the ice sheet had covered Scandinavia, but also that it was land ice. Lyell's theory of drift had led many to think of it as sea ice, or rather a frozen sea; the idea was that the Arctic Ocean, with its gigantic icebergs, had flooded large areas. Torell argued that the primary cause of the ice age had to be sought in a climatic change, namely, a fall in temperature. He also maintained that the ice had eroded and deposited material from the primary rock, which was by no means considered obvious at this time. In reaching his conclusions, he drew much on the observations of his colleagues Hampus von Post and Axel Erdmann. All the loose strata of Sweden had been formed by the action of the ice in tearing off debris from the rocks and grinding it into gravel, sand, and clay. He also considered, unlike Sefström and Lyell, that the striations and the transport of the erratic boulders had been caused by the movements of the ice. In a lecture to the Academy in 1864, Torell stated his view on the erratics. In particular he wished to refute Lyell's hypothesis of a vast ocean of floating icebergs, i.e., the drift theory. He asserted that Sweden and the Continent had instead been joined by a land bridge, and that the land ice had carried the boulders with it.

Torell did not rest on his laurels. Shortly after his dissertation, in May 1859, he was on his way to northern Greenland. Of this expedition we know less, since he wrote no report on it. But we do know that he climbed to the inland ice cap and that he gathered and brought home for analysis a large amount of material. So far Torell had for mostly financed his expeditions himself, but naturally this could not continue indefinitely. In the future he

would need greater financial resources, and the Academy undertook to raise funds and assume responsibility for polar exploration.
Torell began to plan a major expedition to the Arctic, and this
resulted in the first "official" expedition in 1861. We have a detailed
account of it by one of its members, Karl Chydenius, which was
also published in book form.[11] The Riksdag allocated the not in-
considerable sum of 20,000 *riksdaler,* and Prince Oscar (later King
Oscar II) contributed 4,000. The expedition began to become a
national concern.

Torell prepared his expedition most thoroughly, and it was to
set the standard for many later ones. In Copenhagen he recruited
the polar explorer Carl Petersen as a member of the expedition,
and Captain Ammondsen undertook to obtain dogs from Greenland. In London Torell consulted the experienced Leopold
McClintock, who, as we have seen, had finally solved the mystery
of the lost Franklin expedition, and the geologist Roderick I. Mur-
chison, who was now the head of the Geological Survey and president of the Royal Geographical Society.

The expedition had at its disposal 26 men and two Arctic
vessels, *Aeolus* and *Magdalena.* Torell's plan for the enterprise
had two main purposes—the scientific exploration of Spitzbergen
and its coasts, and a geographical excursion farther north. The
experts of the Academy liked this proposal, for it would enable a
lack of success in one sphere to be compensated for in the other.
Geology still had pride of place, with particular attention to be
devoted to the glacial phenomena, but in zoology, botany, and
physics, there were clearly specified subjects to be investigated.

The results of the expedition were hardly earth-shattering, but
it yielded a considerable volume of both specimens and reports.
Torell and Nordenskiöld collected minerals and examined glaciers, and found further confirmation of Torell's theories. In the
introduction to his book, Chydenius makes the point that the
program of the expedition was scientifically based. Previous voyages of exploration towards the North Pole, by the British, the
Americans, and the French, had been given a new direction with
the "discovery of the ice age," he writes.

Nordenskiöld as an Arctic Explorer

His participation in Torell's expeditions had given A.E. Nordenskiöld valuable experience of polar exploration. He seems also

to have become more and more fascinated by the subject. Another expedition to Spitzbergen was organized in 1864; the Riksdag again voted funds; and Nordenskiöld was now put in charge. The primary purpose was to complete the measurements of a degree of the meridian that it had not been possible to carry out in 1861. He was accompanied on the schooner *Axel Thordsen* by the physicist Nils Dunér and the zoologist Anders Johan Malmgren; they had a crew of three men. After taking many astronomical fixes, Nordenskiöld and Dunér were able to produce a map of Spitzbergen far more reliable than any previous one.

The expedition of 1864 marked the end of the first phase of Swedish polar exploration. The voyages to Spitzbergen had all been made in sailing ships; henceforth, with a single exception, steamers would be used. The overall result of the expeditions that had been carried out was very satisfactory. Apart from the cartographic work and the diaries, 34 scientific works were published, dealing with geography and terrestrial magnetism, geology and palaeontology, botany and zoology. "As far as some aspects of its natural history are concerned, Spitzbergen may now be reckoned among the most thoroughly investigated countries in the world," wrote Malmgren.[12]

However, the Swedes also had another ambition: to penetrate as far north as possible. This resulted in the expedition of 1868, which was financed by private donors, foremost among them the merchant Oscar Dickson in Gothenburg. The expedition reached latitude 81° 42′ N, which was a record for a ship. It also brought back large quantities of specimens and reports of observations. Nordenskiöld admitted that the idea of an open polar sea was contradicted by experience and that the only way to reach the pole would be to spend the winter in the Arctic before continuing northwards on the ice by sledge in the spring. He at once began to plan such an expedition, again receiving financial backing from Dickson, and he himself led a reconnaissance to Greenland in 1870; this developed into a full-scale expedition that produced large geological collections. Nordenskiöld brought home, for example, large lumps of iron, which he thought to be of cosmic origin; however, they have since been found to be tellurian. The wintering expedition itself was launched in 1872–1873 and met with severe setbacks. Ice conditions made the drive for the north impossible; the reindeer that were used as beasts of burden in preference to dogs could not stand up to the ordeal and died; provisions nearly ran out; and scurvy was rife among the crew.

Nevertheless, Nordenskiöld had gained further valuable experience from his treks across the ice, and he now began to turn his attention to the Arctic Ocean north of Siberia. His next goal was to find the North-East Passage. This major discovery is far too well known to need to be described in detail here. The Academy of Sciences, which was also, as we know, Nordenskiöld's employer, was still where his planning was co-ordinated. Here he received scientific assistance from his colleagues; here he presented the results of his expeditions; and here he could discuss every conceivable problem. After two preparatory expeditions to the Yenisei in 1875 and 1876, work began on organizing the big one. The costs were borne mainly by Oscar Dickson and the Russian mineowner Alexander Sibiriakoff, but substantial grants were also made by the Riksdag and by King Oscar II. The seal-fishing vessel *Vega* was bought and fitted out for its special mission; the experienced Lieutenant Louis Palander was put in command; and a large crew was assembled, including not only Swedes, but also an Italian, a Dane, and a Russian.[13]

In July 1878, the expedition set off from Tromsø on the north coast of Norway, continuing through the sound between the mainland and Novaya Zemlya and across the Kara Sea to the mouth of the Yenisei. The journey continued to make good speed all the way to the entrance to the Bering Strait, where winter quarters had to be set up at the end of September. Throughout the winter the group carried out scientific surveys, but they also had time to make contact with some of the inhabitants of the region. By July 1879 the ice had broken, and the *Vega* was able to resume her journey through the Bering Strait. With this the North-East Passage had been completed. The rest of the journey was a triumphal voyage of unheard-of proportions. Nordenskiöld and his men were fêted with receptions of honor and state banquets throughout their journey home, and when they reached Stockholm in April 1880, enthusiasm knew no bounds. The King and the people of Sweden paid homage to their achievement. Nordenskiöld was made a baron, and Palander was ennobled with the name Palander of Vega.

Polar Exploration as a Fashionable Science

Thus Nordenskiöld became a hero. The project that had begun as a purely scientific activity had become a source of na-

The well-known portrait of scientist and heroic explorer A.E. Norden-skiöld on the Arctic ice. Painted by Georg von Rosen in 1886. (Courtesy of Nationalmuseum, Stockholm.)

tional pride. It is, of course, difficult to pinpoint when it began to assume this character, but it seems obvious that the private patrons played an important part in it. That the King and the Riksdag became involved at an early stage naturally enhanced the status of the enterprise, but the large private investment was even more significant. Nordenskiöld himself said that if Dickson had not given such whole-hearted backing, Swedish polar exploration would have come to a halt after the expedition of 1864.

Oscar Dickson was to pay for six expeditions, either in whole or in part. We may wonder why. There was a long tradition of support for art and science in Gothenburg, going back to the days of the East India Company. Anyone who had become rich was expected to be generous to the public, by founding a hospital or a school, or by paying for scientific expenses. Dickson chose polar exploration, and he did it in a big way. He not only funded Nordenskiöld's expeditions, he later continued to support Fridtjof Nansen, S.A. Andrée, and Otto Nordenskjöld (nephew of Adolf Erik). The ostensible motive was to serve his country. Nor has anybody questioned his good intentions. But he knew, of course, that he would also gain a good reputation and be well rewarded. His lumber operations, particularly along the coast of Norrland, had made him a multimillionaire, and his name is associated with illegal overcutting, leading to lawsuits. His scientific patronage enabled Dickson to earn a reputation of a different sort. He became in rapid succession a member of the Royal Geographical Society of London, an honorary doctor of philosophy when Uppsala celebrated its quartercentenary, a member of the Academy of Sciences, and an honorary member of the Royal Society of Naval Sciences; he was ennobled, received medals of honor, and was finally made a baron.

Interests thus coincided in a most satisfactory manner. Dickson supported science and his own country, and received a good dividend on his investment. He also got his portrait in Nordenskiöld's book on the voyage of the *Vega*. [14]

How, then, did Nordenskiöld see himself and his efforts? Everything points to his having been a realist. His writings are highly factual; he never launches into flights of fancy on his heroic exploits. He starts off his books by describing the background to the expedition and its equipment. The last page usually lists the results. We may sometimes feel that the pages might have been allowed to reveal a little more of the drama, but Nordenskiöld wastes no space on things like that.

This does not, of course, mean that he was unaware of the aura that surrounded the polar voyages, but he did not wish to jeopardize his scientific reputation. Naturally he was fully aware of the financial and patriotic aspects of the picture; he well understood the heroic role that had been assigned to him and of which he had had such vivid experience on the voyage home with the *Vega*. But he was careful to distinguish between science and the trappings that surrounded it. In the foreword to the report on the Yenisei expedition of 1876, he stresses that this is a scientific expedition, "not a commercial enterprise."[15] But sometimes Nordenskiöld was obliged to blow his own trumpet, particularly when trying to mount a new expedition. In introducing his planned voyage with the *Vega*, he mentions the wealth of scientific material he has brought home to the Museum of Natural History, making it the richest such museum in the world:

> "Then there are the discoveries and the surveys which promise to become, or have already become, of practical significance, e.g., the meteorological and hydrographic work of the expeditions, their extensive study of seal and walrus catches in the Arctic, the demonstration of hitherto unsuspected stocks of fish along the coasts of Spitzbergen, the discovery of the valuable coal and phosphate deposits of Bear Island and Spitzbergen itself, which ought in future to become of great economic importance to countries in the region, and above all the successful penetration to the mouth of the great rivers Ob and Yenisei, navigable to the very borders of China, which has solved a centuries-old navigational problem."[16]

However, Nordenskiöld prefers to concentrate on the scientific expectations. He speaks enthusiastically of the meteorological and geological results; he expands on what may be gained from the study of plant and animal life in the polar regions; he describes the interest in "the discovery of colossal elephantine remains in Siberia's frozen soil" and tells of "semipetrified or carbonized vegetable remains from several different geological epochs." These are as yet unconfirmed, says Nordenskiöld, but they should be enough to permit the expedition to be seen in the same light as similar ventures that have "set off from these shores and brought benefit to science and honour to the name of Sweden." An expedition such as the one now planned will also benefit Swedish seafarers in general, "by the increase in self-respect that an awareness of the loyalty, courage and resilience of comrades always brings."

Nordenskiöld was sober-minded and shrewd, but he played his part in making polar exploration a fashionable science. He lived in an age with a strong faith in the future. Auguste Comte and Herbert Spencer forecast the improvement of man and society; Thomas Henry Buckle saw the history of civilization as a tale of unbroken progress. The great world exhibitions, starting in London in 1851, were intended to demonstrate the achievements of science and the blessings of industry. The aim was, wrote a British newspaper, "to seize the living scroll of human progress, inscribed with every successive conquest of man's intellect."[17] The great voyages of discovery have to be seen as part of this pattern, whether their destination was Africa or the North Pole. As well as economic, national, or scientific motives, there were, of course, individual ones: man's striving to conquer nature was symbolized by the struggle of the explorer-scientist in unknown lands. Optimism and the belief in progress had found their heroes.

As an example of the fascination exerted by the polar expeditions, we may look at the works of an adventure writer such as Jules Verne. He had an eye for the exotic and the exciting, and he produced a large work on the history of geographical discoveries, which shows that he was no mere dilettante.[18] The polar region became one of his favorite subjects. Five years before Nordenskiöld wintered in the Arctic, Verne published a story entitled *A Winter on the Polar Ice.* He had his Captain Nemo conquer the South Pole long before Amundsen and Scott got so far in reality. And he wrote a novel on the voyage of Captain Hatteras to the North Pole, again before anyone had reached this magic point on the earth's surface.

Now, it may well seem a big step from the fantasies of Jules Verne to the highly concrete research projects of Nordenskiöld. But although Nordenskiöld was, of course, a sound scientist, he did not remain unaffected by the contemporary view of the scientist as a hero. He understood his role and knew how to appreciate the description of it in literature. When Jules Verne's story of Captain Hatteras came out in a cheap Swedish edition in 1892, Nordenskiöld himself wrote a handsome preface for it. Our time, he says, is characterized above all by the impatient and furious desire of the civilized peoples to reveal the secrets of nature, to explore all the countries of the earth, "to make all the forces of nature the servants of man's will."[19] Bolder than ever, biologists, chemists, physicists, and engineers of every description compete to "invent new methods of taming nature and turning its force to man's advantage as he wishes." Thanks to new instruments, Nordenskiöld goes on, nothing less than a complete transformation of our knowledge of the cosmos is taking place. As well as technical literature, a stream of popular writings is appearing, to "spread knowledge of all the new and wonderful things that the scientist has thus discovered, devised or suspected" to the general public. Among such writings the books of Jules Verne take "their own particularly prominent place." To challenge and seek to control nature thus became part of the philosophy even of such a scientist as Nordenskiöld.

The Period after Nordenskiöld

The triumph of Nordenskiöld and the general enthusiasm beyond scientific circles created a favorable atmosphere for continued polar exploration. Nordenskiöld himself went on only one more expedition, to Greenland in 1883, but from his position in the Academy, he advised others. Important decisions on expeditions were hardly ever taken before he had been consulted.

In the decades that followed Nordenskiöld's successes, polar exploration fell into one of two categories. One was a continuation of the purely scientific research that had been begun by Torell, i.e., meticulously prepared voyages for cartography, observation, and other detailed studies. These expeditions were characterized by their scientific breadth, since several different scientific disciplines were usually represented. This research was thus a matter of hard work rather than heroic deeds. The second type of project gave prominence to the element of adventure implicit in polar exploration. Many wished to be heroes in Nordenskiöld's footsteps. Fridtjof Nansen was directly inspired by his Swedish predecessor, and was himself a combination of scientist and adventurer. He stated the glacial theory as the source of his interest in the polar regions when he embarked on his voyage to Greenland in 1888.[20] His grand idea of the drift of the pack ice was the theory of a scientist, but his testing of it was the action of an adventurer. Nansen had read that pieces of the wreck of the American vessel *Jeanette,* which had sunk off Siberia in 1879, had been found two years later in Greenland. He therefore reasoned that the drift ice

Wintering on the *Vega*. Illustration from Nordenskiöld's own book of his voyage through the North-East Passage.

would carry a frozen ship from north-east to south-west right across the North Pole.[21] He had the *Fram* built to test this theory and let it become icebound, whereupon the ice carried it off as predicted, even if, unluckily for Nansen, it did not cross the actual pole. Nansen took the view that this point was not really that important in itself; the essential thing was the accompanying series of scientific observations. At the same time, he stated his adventurer's philosophy concerning man's fight against the elements: "Nowhere, surely, has knowledge been purchased at such cost in deprivation, need and suffering; but the human spirit will not rest until every spot, even in these regions, has been trodden by human foot and every puzzle up here has been solved."[22]

S.A. Andrée was also a scientist, but the heroic role got the better of him. His idea of crossing the North Pole in a balloon won Nordenskiöld's approval; otherwise Andrée would never, he said, have dared put it into practice. The King backed the project, and Oscar Dickson was prepared, as on previous occasions, to underwrite it financially. The meteorologist Nils Ekholm was to take part in the expedition, but withdrew when he decided the balloon was unsatisfactory in quality and durability. Andrée himself was not critical enough, and this had fateful consequences. The three members of the expedition, Andrée, Knut Fraenkel, and Nils Strindberg, left Spitzbergen with their balloon *Örnen* (The Eagle) in July 1897 and vanished without trace. Not until 1930 were the expedition's last camp and the remains of the three men found on White Island, north-east of Spitzbergen. Preserved diaries and rolls of film enabled the story of the failure to be reconstructed in detail. Soon after it started the balloon had been weighed down by ice and landed. The three men had to struggle on across the ice, but as this was constantly in motion they drifted in the wrong direction and failed to reach land in time. The actual cause of their deaths was probably trichinosis, after eating badly cooked polar bear meat.[23]

The disappearance of the Andrée expedition undoubtedly dampened the spirit of adventure in Sweden, but abroad many sought to outdo one another in the role of hero. The race between Amundsen and Scott for the South Pole has in retrospect come to appear in a faintly ridiculous light. And the pathetic attempt of Robert E. Peary to reach the North Pole, to be "the first," failed to convince even at the time. He was never really recognized as the winner, and nowadays it is regarded as certain that he was bluffing. He never reached the Pole.

The more serious, purely scientific side of research received less publicity, but the level of activity was impressive. In the years up to 1910, 35 Swedish expeditions set off for the polar regions (25 "major expeditions" and ten less ambitious ones).[24] Beside the usual scientific surveys, there was an increasing interest in finding coal. In 1896 a Swedish expedition led by Gerard De Geer surveyed the whole area around Isfjorden and its coal deposits. In the summers of 1898 and 1899, Alfred Nathorst led expeditions with the secondary objective of finding Andrée and his men. In the first year, important cartographic work was done, and geological and biological surveys were made of Bear Island and the previously almost unknown King Karl Land. For the first time, Spitzbergen was circumnavigated by a Swedish vessel. In the second year, the destination was north-east Greenland, which led to the discovery of King Oscar's Fiord and the hitherto unknown Devonian formation. The year 1899 saw the start of Swedish-Russian collaboration in expeditions to measure a degree of the meridian, the full project stretching over five summers and including one winter in the Arctic. This enabled the flattening of the earth at the poles to be more accurately described. At about the same time, Norwegian concessions were granted for coal mining on Spitzbergen at Advent Bay (part of Isfjorden); these were later sold to British and American companies.

There had been some mining operations on Spitzbergen in 1906, and the capital, Longyear City (later Longyearbyen) had been founded. (J.M. Longyear was the head of an American company that minded coal on Spitzbergen between 1904–1916.) In 1910 the Swedes launched a prospecting expedition and succeeded in finding coal at several places in Isfjorden. In the following year, a company was formed to exploit the deposits. At that time the world market price of coal was too low, however, but after the outbreak of war, prices rose quickly, and a new Swedish company was formed. In 1917 *Sveagruvan* (the Svea mine) was sunk and operations started. During the following years, a maximum of 216 persons spent the winter on Spitzbergen, all employed in various capacities in mining. Around 1920 the price of coal dropped again, and the Swedish government had to take over the financial liability. In 1925 a fire broke out at the mine and proved difficult to extinguish. Operations had to be suspended, and the company went into liquidation. After a few years, it was reconstructed and sold to a Norwegian concern.

The coalmining operations on Spitzbergen gave rise to international complications and demands for a definite status for the area in international law. Several conferences were held without result, but in 1920 the Treaty of Svalbard gave Norway sovereignty over the archipelago on special conditions: these included the demilitarization of the area and a free right for all countries to exploit its natural resources. Most of the subsequent research has been carried out by Norway, from 1928 by Norges Svalbard- og Ishavsundersøkelser and since 1948 by the Norwegian Polar Institute.

For the sake of completeness, it should be mentioned that Sweden has also taken an interest in the South Pole. The driving force here was Otto Nordenskjöld, mineralogist and geologist, who was undoubtedly inspired by his famous uncle. Otto Nordenskjöld had already visited both the Arctic and the Antarctic when he began to plan a major expedition to the Antarctic in 1901, which was in part expected to coincide and collaborate with German and British expeditions. The expedition took two years but found itself in desperate straits when its vessel, the *Antarctic,* foundered. However, the whole party survived, and a relief expedition led by Olof Gyldén succeeded in effecting a rescue. Nordenskjöld became professor of "geography and ethnography" at Gothenburg in 1905 and later made further expeditions; in the summer of 1909, he visited Greenland, and in 1920–1921 he explored the ice massif of southern Patagonia in Chile. From 1908–1918 he sat as one of Sweden's two representatives on the International Polar Commission, becoming its chairman in 1913. Nordenskjöld was a member of the Academy of Sciences and also became known as a prolific author, his writings including many works of popular science. Between 1911–1914 plans reached an advanced state of preparation for a Swedish-British expedition to establish a permanent base in Antarctica. By 1914 Otto Nordenskjöld and Louis Palander had foundation drawings ready, and the Swedish Government had promised a financial subsidy. At this point the First World War broke out, and plans had to be abandoned.[25]

Modern Research

Since the failure of its mining operations in the 1920s, Sweden's interest in the polar regions has been concentrated on scientific questions. One of the most important names is that of Hans W:son Ahlmann, professor of geography at Stockholm University from 1929 onwards and also an active member of the Academy for many years; his extensive collections are preserved in the Academy, although not yet examined and analysed.[26]

Between 1920 and 1925, Ahlmann made a number of visits to Jotunheimen in Norway to study the glaciers there. He led one large Swedish-Norwegian expedition to Spitzbergen in 1931 and another (together with Harald Sverdrup) in 1934. A few years later, in 1936, he and Jon Eythórsson led a Swedish-Icelandic expedition to explore Vatnajökull, in Iceland, the biggest glacier in Europe. Another major expedition, this time to north-east Greenland, was led by Ahlmann in 1939.

Ahlmann's importance lies not only in his own research, but also in the teaching and encouragement he gave to a new generation of polar scientists, several of whom became leaders in the field. It was also on his initiative that a large Norwegian-British-Swedish expedition went to Queen Maud Land in the Antarctic in 1949–1952.

Sweden participated in the second International Polar Year, 1932–1933, when two stations were built on Spitzbergen for meteorological and magnetic observations. During the third polar year, known as the International Geophysical Year 1957–1958, Sweden was represented on Spitzbergen in a Swedish-Finnish-Swiss expedition to Nordlandet under Professor Gösta Liljequist. The program included geophysics and meteorology, atmospheric chemistry, atmospheric electricity, cosmic radiation, the aurora borealis, terrestrial magnetism, and tides. As glaciology was not represented, the expedition was supplemented with a separate glaciological working party led by Valter Schytt. The glaciologists took the same route over the inland ice in 1958 as Ahlmann had taken in 1931, and they were able to supplement and correct his results. An American member, Weston Blake Jr., examined the shorelines, which showed that the land had previously been depressed by thick land ice. During the 1960s a series of summer expeditions to Spitzbergen was made by researchers from the department of physical geography at Stockholm University and led by Gunnar Hoppe.

Next came the Nordenskiöld Centenary of 1978–1980. Plans were laid in good time for a polar expedition from Sweden to mark this anniversary. The chief organizers were the Academy of Sciences and the Swedish Society of Anthropology and Geogra-

The ice breaker *Ymer* was used for the Nordenskiöld centenary expedition to the Arctic in 1980. Drawing by Gunnar Brusewitz.

phy, but valuable assistance was also given by the National Administration of Shipping and Navigation and the commander-in-chief
of the Royal Swedish Navy, Vice-Admiral Bengt Lundvall. The
idea was to sail the full North-East Passage, but this required
Soviet permission; it was proposed that the expedition should be
a joint Swedish-Soviet one. However, political obstacles arose, and
in the end an all-Swedish Arctic expedition with rather more limited objectives took place. The Swedish government was wholly in
favor of the venture, and the expedition was allowed to use the
government ice-breaker *Ymer.* There was also collaboration with
the Norwegian Polar Institute. The expedition concentrated its
efforts on clearly specified research projects: atmospheric chemistry (especially air pollution in Arctic regions), glacial meteorology
(the importance of clouds of mist in the radiation balance), oceanography (including movement, rate of flow, and circulation of the
water), marine biology (how different species evolve in the polar
ocean), terrestrial biology (how plants and animals can survive in
the severe polar climate), geology, and physical geography (investigation of shorelines, examination of the bottom sediments of the
Barents Sea, and collecting samples from the deep-sea bed). Altogether more than 100 scientists from nine different countries took
part in the Ymer-80 Expedition, which operated in an area from
Franz Josef Land in the east to north-east Greenland in the west.[27]

In the 1980s the Academy of Sciences has again found itself at
the center of polar research. Following a decision of the Riksdag
in 1984, a Swedish Polar Research Secretariat has been set up, with
its headquarters at the Academy. The Secretariat is a governmental organization under the Ministry of Education. Its task is to
organize and supervise Swedish polar research activities in the
Arctic and Antarctic regions. The Secretariat works in close co-
operation with the Polar Research Committee of the Academy of
Sciences, which has the status of a national committee. The Polar
Secretariat has now organized the biggest Swedish expedition yet
sent to the Antarctic, under the name Swedarp 1988–1989. Its
work involves both scientific surveys and establishing a permanent research station in Queen Maud Land. In its program the
Secretariat has explained Swedish interest as follows:

"Swedish interest in the polar regions has an obvious basis
in Sweden's geographical proximity to the North Pole.
Furthermore, the Swedish landscape was formed by processes which can be studied in Greenland and the Ant-

arctic. The subarctic environment and the severe climatic conditions in the north of Sweden create a potential and a need for polar competence not only in science but also in technology and in the many practical aspects of every-day life. With this in mind Sweden now has embarked upon an ambitious development of its polar activities in general and polar research in particular."[28]

This program shows that Swedish polar research still rests firmly on scientific foundations, even if external circumstances have changed. Geopolitical considerations and the demands of modern science for expensive equipment have naturally affected the conditions for research. Small private expeditions such as To-rell's are nowadays hardly possible; instead there are international agreements on collaboration and on major jointly organized expe-ditions. It should also be mentioned here that Sweden reached an important political and scientific goal in September 1988, when it became a consulting member of the Antarctic Treaty and a full member of the Scientific Commission for Antarctic Research (SCAR). So there is indeed an ambition to participate in this excit-ing area of research.

Notes

1. This paper is partly built on earlier studies, especially *Upptäckten av istiden* (Uppsala, 1976), which deals with the growth of modern Swed-ish geology.
2. See, for instance, L. P. Kirwan, *A History of Polar Expeditions* (New York, 1960). Sir John Franklin is the object of a novel by Sten Nadolny, *Die Entdeckung der Langsamkeit* (München, 1983).
3. Cf. my article on Nordenskiöld and the theme on the scientist as a hero in *Vetenskapsmannen som hjälte* (Stockholm, 1984).
4. There are many good surveys of the first stage of the history of modern geology, e.g., Roy Porter, *The Making of Geology: Earth Science in Britain, 1660–1815* (Cambridge, 1977); Martin Guntau, *Die Genesis der Geologie als Wissenschaft* (Berlin, 1984); Rachel Laudan, From *Mineralogy to Geology: The Foundations of a Science, 1650–1830* (Chicago, 1987). On fossils, see Martin J. S. Rudwick, *The Mean-ing of Fossils* (London, 1972).
5. Laudan, ch. 9. Cf. Roy Porter, "Charles Lyell and the Principles of

the History of Geology," *British Journal for the History of Science* 9 (1976), 91–103. For the discussion on uniformity, see R. Hooykaas, *Natural Law and Divine Miracle* (Leiden, 1959), and Rudwick, "The Principle of Uniformity," *History of Science* 2 (1962), 82–86.

6. Cf. my article "The Geological Ideas of J. J. Berzelius," *British Journal for the History of Science* 9 (1976), 228–236.
7. Rudwick, "The Glacial Theory," *History of Science* 8 (1969), 136–157.
8. Frängsmyr, *Upptäckten av istiden*, ch. VI. *Yoldia arctica* should be *Portlandia arctica* (Gray), also called *Nucula arctica;* see my edition of Torell's *Travel Book* 1862–1863, "Otto Torells reseberättelse 1862–1863," *Lychnos* 1986, 127–148.
9. Otto Torell, "Undersökningar öfver istiden" I, *KVAö* 1872, 29.
10. Torell, *Bidrag till Spitsbergens molluskfauna, jemte en allmän öfversigt af arktiska regionens naturförhållanden och forntida utbredning* (Lund, 1859).
11. Karl Chydenius, *Svenska expeditionen till Spetsbergen år 1861, under ledning af Otto Torell* (Stockholm, 1865).
12. Quoted by Nathorst, "Svenskarnes arbeten på Spetsbergen," i *Nordisk Tidskrift* 1906, 174.
13. George Kish, *North-East Passage: Adolf Erik Nordenskiöld, His Life and Times* (Amsterdam, 1973). A special issue of *Ymer* 1902 was dedicated to "the memory of A. E. Nordenskiöld," containing six articles; among them A. G. Nathorst, "Nordenskiölds polarfärder," 141–206, and *idem.*, "Nordenskiöld som geolog," 207–224. See also *Ymer* 1980, "Vega 1880–Ymer 1980."
14. Nordenskiöld, *Vegas färd kring Asien och Europa*, 2 vols. (Stockholm, 1880–1881). Cf. my article "Nordenskiöld och polarforskningen," *Ymer* 1980, 35–36.
15. Nordenskiöld and Hj. Théel, *Redogörelser för de svenska expeditionerna till mynningen af Jenisej år 1876* (Stockholm, 1877), förordet.
16. Nordenskiöld, *Framställning rörande 1878 års ishafsfärd* (Göteborg, 1877); eng. transl. *Memorial concerning the arctic expedition of 1878* (Göteborg, 1877), 30 pages.
17. Quotation from J. B. Bury, *The Idea of Progress: An Inquiry into its Growth and Origin* (1920; New York, 1960), 329.
18. Jules Verne is treated in my book on the utopian tradition, *Framsteg eller förfall* (Stockholm, 1980), 199–220. Jules Verne's history of geographical discoveries in three large volumes was translated into Swedish, 1880–1884.
19. Nordenskiöld, "Företal," in Jules Verne, *Kapten Hatteras' reseäfventyr,* 2 vols., new ed. (Stockholm, 1892).
20. Fritjof Nansen, *På skidor öfver Grönland* (Stockholm, 1890), preface.
21. Nansen, *Fram öfver Polarhafvet*, 2 vols. (Stockholm, 1897), translated by A. G. Nathorst, introduction.

22. *Ibid.*, 2 f.
23. Per Olof Sundman, *Ingen fruktan, intet hopp: Ett collage kring S. A. Andrée, hans följeslagare och hans polar-expedition* (Stockholm, 1968); Sven Lundström, *Andrées polarexpedition* (Höganäs, 1988).
24. A. G. Nathorst, et al., "Swedish Explorations in Spitzbergen 1758–1980," *Ymer* 1909. I have also used Urban Jonsson, "Svensk polarforskning på Spetsbergen 1858–1902" (manuscript). Cf. J. M. Hulth, "Swedish Arctic and Antarctic Explorations 1758–1910," *KVAA* 1910, bilaga 2.
25. Gösta H. Liljequist, "Swedish Antarctic Research—A Historical Review," in *Sweden and Antarctica*, ed. Anders Karlqvist (Stockholm, 1985), 7–58. (Bibliography, 89–93.)
26. Cf. Staffan Helmfrid, "Hans W:son Ahlmann 1889–1974," *Zeitschrift für Gletcherkunde und Glazialgeologie*, Vol. X, 249–257.
27. A report consisting of 22 papers is given in *Expedition Ymer-80*, published as the yearbook *Ymer* 1981 (Stockholm, 1981).
28. "Swedish Polar Research," program from the Swedish Polar Research Secretariat.

BOSSE SUNDIN

Environmental Protection
and the National Parks

THE commitment of the Academy of Sciences to questions concerning the protection and preservation of the environment and
natural resources can be said to be based on traditions reaching
back to the Academy's very beginnings. During the eightteenth
century, for example, the Academy devoted much time to forestry
questions (one of the burning issues of the day, in that it was felt
that the life-giving forests of Sweden were on the verge of deple-
tion). But the secretary of the Academy, Christopher Aurivillius,
was correct when, in his annual report of 1910, he observed that
new and important tasks were at hand, when the Academy was
assigned the ultimate responsibility for realizing the intentions
behind the Protection of Nature Law passed by parliament the
preceding year. This was Sweden's first legislation dealing with
natural landmarks and national parks, for which ten areas had
been proposed. Among the new duties of the Academy, were
managing these new national parks, and providing expertise in the
protection of natural landmarks.

It was impossible, said Aurivillius, to predict the scale on
which this work would be carried out, but questions pertaining to
the protection of nature would probably demand considerable
attention in the future, and might indeed prove to be the heaviest
of all the Academy's responsibilities.[1] Time has shown him to be
if not completely, then at least partly, prophetic. Such questions
developed into one of the Academy's most important tasks, and its
Committee for the Protection of Nature was a center for Swedish

environmental protection activities throughout the first half of the twentieth century.

Today, 80 years later, engagement in the protection and preservation of the environment continues to play a central, indeed growing, role in the work of the Academy, although the duties themselves have changed. Government agencies such as the National Environmental Protection Board have taken on the Academy's earlier management and administrative responsibilities; instead, purely scientific tasks have come to dominate. Even the questions themselves have changed. No one speaks of "the protection of nature" any longer; today, the discussions are dominated by environmental, natural resource, and energy questions. The Committee for the Protection of Nature has been succeeded by an Environmental Committee, and protection of Sweden's nature and natural landmarks is no longer the ultimate goal. Instead, matters of global concern are discussed, questions involving the very survival of mankind. However, there was a long road to this stage of awareness, and a discussion of the engagement of the Academy in environmental and natural resource questions must begin at the turn of this century, with the circumstances preceding the birth of organized environmental protection in Sweden.

The Birth of Environmental Protection

In the late 1870s and early 1880s, the first voices were raised calling for environmental protection in Sweden. The most prominent among these pioneers was Adolf Erik Nordenskiöld who in 1880 (the year he returned triumphantly with Vega after having discovered the North-East Passage) proposed the establishment of "protected national parks" in the Nordic countries. Progress, he stated, had occasioned such interference with nature that in the future it would be difficult to gain an understanding of that nature "with which our ancestors fought their first battles." Still, there remained distant regions where the soil and the forest were of little commercial value, and where nature had been preserved in its pristine condition. There, without any major sacrifice, a suitable area could be chosen and declared a national park, "where forest and land and lake could remain untouched, where trees could not be felled, thicket not cleared, grass be left unmown, and where all animals who were not actually vermin could be safe from the hunter's rifle all the year round."[2]

Nordenskiöld's proposal, most likely inspired by the first national parks in the United States, soon became a topic for lively discussion. Though no immediate measures resulted, the national park idea doubtless stimulated many scholars to reflect on where and how such a plan might be implemented.

By the turn of the century, the nature protection idea had won many supporters.[3] Most of the first advocates of this idea in Sweden were—not surprisingly—active natural scientists, principally botanists, zoologists, and geologists, associated with the universities in Lund and Uppsala and with the Academy of Sciences and the Museum of Natural History. But other groups also rallied round the banner of the protection of nature. For example, there were the artists and writers of the "national romantic" spirit, who expressed a new appreciation for the emotional, aesthetic, and economic value of Swedish nature. Hunters and representatives of the forestry industry could for more practical reasons advocate making more rational use of natural resources; in fact, hunters, alarmed by the threatened extinction of game, were among the first to express concern about the changes in the natural environment. Another characteristic feature of early environmental awareness was the striving to protect insectivorous birds, presumed to be of economic importance for agriculture and forestry, since they held noxious insects in check. Finally, we should mention the growth of tourism, organized in 1885 under the umbrella of the Swedish Touring Club, (*Svenska Turistföreningen,* STF), which wanted portions of the Swedish countryside exempted from immediate economic exploitation. The world's first national parks, e.g., Yellowstone National Park in the United States and Rocky Mountains National Park in Canada, had also been established with the clear goal of serving as "amusement parks" for tourists.

The varied groups mentioned above also reflect the motives behind the protection of nature in the early twentieth century. In a nature protection report from 1907, the arguments in favor of protection are summarized in a typical manner. First, the *economic motive* is named. It was necessary to restrain "heedless Avarice" from helping itself to the riches of nature in a manner that would endanger future supplies. To this end, there already existed some legislation, such as hunting and fishing regulations or forestry legislation. Now, however, other motives had arisen—above all, it was stated, the *scientific motive*. It was of paramount importance that science be enabled to follow the evolution of nature within specific areas where it had been left entirely to itself, and that rare animal and plant life and other noteworthy natural phenomena be protected against destruction. And indeed, not just scientists were interested in primeval nature—for there was also the *aesthetic motive*. A natural object could contain an element of beauty that could awaken a love of nature among the general populace. Finally, the *historical or cultural motive* was mentioned: certain natural objects were worth saving because of the legends, historical memories, or folk traditions associated with them.[4]

In retrospect one might add a further motive that acted with and reinforced both the scientific and the aesthetic and cultural-historical motives: the *nationalistic motive*. The protection of nature movement was inspired by a newly awoken patriotism that came into being around the turn of the century. For the older, nineteenth-century patriotism, Sweden's greatness was to be found above all in its proud history, studded with military victories. The new national enthusiasm was characterized by a biologically influenced conception of the nation in which nature provided its distinctive element. The moors, forests, and other natural wonders became national symbols with a deeper significance than that of the heroes and warrior-kings of the past. Scientific inventory and researching of the country's natural resources undoubtably played a large part in emphasizing these new national symbols.

Yet the decisive impulse for organized nature protection came from without. In early 1904 Sweden played host to Professor Hugo Conwentz, the foremost spokesman for nature protection in Germany. Conwentz lectured in Stockholm, Uppsala, Gothenburg, and Lund on the dangers threatening the natural landscape and its plant and animal populations. These talks, which were widely publicized were the primary impetus for a motion in parliament by Karl Starbäck proposing an inquiry into appropriate measures for protecting Sweden's nature and natural landmarks.

Starbäck, a botanist with a Ph.D. from the University of Uppsala, pointed out how other nations—mainly the United States and Germany—had realized the fundamental importance of protecting nature. While changes in the natural environment in Sweden certainly had not gone as far as in many other civilized countries, it was still difficult to find areas untouched by man. Therefore, according to Starbäck, it was high time to act by following the examples set abroad and take steps to protect the nature and natural landmarks of Sweden.[5]

Starbäck's proposal, which won parliamentary support, was already firmly established in the Academy of Sciences. The geologist Gerard De Geer, member both of parliament and the Academy, let the motion itself record his full agreement with its aim. In an appendix, nine more scholars declared that they too supported the motion. Among these were secretary of the Academy Aurivillius and four other members of the Academy.

The Academy of Sciences was given the task of executing the proposed inquiry. In September 1904 the first Committee for the Protection of Nature (as it was subsequently called) was appointed. It consisted of Gustaf Retzius, A.G. Nathorst, Gerhard Holm, Gerard De Geer, and Einar Lönnberg. In their report, which was accepted by the Academy, the committee stressed first the importance of educating the public.[6] Lessons in the elementary schools and institutes of higher learning should be characterized to a greater extent than before by willingness to care for and protect the country's nature. For instance, teachers must remember to stress that the protection of a rare plant or animal in its natural habitat was of far greater value than gathering specimens for some collection or museum. This touches on a problem that constantly arose in that era's discussions about the protection of nature, one that directly concerned the activities of the Academy and its membership. The rarer a plant or animal species was, the greater the risk of its extinction, and thus the more sought after it became for collectors—whether these were private individuals or representatives of a museum of natural history, an institute of higher learning, or any other institution. Therefore, it was of the utmost importance that the lucrative market for natural rarities be dealt with. It may seem strange that this problem first came to the fore when discussion concerned measures for protecting the natural landscape of the nation. But as far as can be adduced, the problem was so broad that it was vital to check this misdirected interest in nature's wonders before it was too late.

Another basis on which the committee's report was built was provided by answers given to a questionnaire sent out to (among others) members of the Academy, a variety of scientific societies, and all superintendents of crown lands and forest officers in Sweden. The committee could thereby produce a catalogue of the most important areas and natural phenomena under consideration for protected status as national parks or natural landmarks. The catalogue was arranged according to province, an indication of the committee's ambition to ensure that the natural types of each part of the country were represented among the national parks.

Even though the Committee for the Protection of Nature referred to certain areas as especially worthy of protection—primarily Gotska Sandön in the Baltic Sea and the area around Stora Sjöfallet in the mountains of Lapland—it preferred to present the catalogue as merely a proposal. It would be up to the National Board of Crown Forests and Lands to conduct the final selection of national parks. According to the proposal, the board should continue to register further restricted areas and even "individual trees." The registration of geological and geographical objects and rare breeds of plant and animal life would be the responsibilities of the Swedish Geological Survey and the Academy of Sciences, respectively. It was also thought that the Academy should take an active role in the protection of rare plant life by appealing to the owners of properties on which threatened species flourished. Furthermore, it was proposed that a bill be drafted whereby individual landowners would be empowered to protect any natural landmark on their property.

Two species threatened with extinction, the bear and the red deer, were given special attention in the report, and the committee suggested that these animals be protected in certain parts of the country. It also recommended that a number of bird sanctuaries be protected.

As was often the case in the early years of the nature protection debate, modern large-scale industry was noticeably absent when the forces threatening nature were identified. When industry was referred to at all, it was from an aesthetic perspective. Magnificent natural scenery must not be destroyed by the devastation of forest land, quarrying, and industrial construction. Instead, it was primarily agriculture that, not entirely without reason, was found guilty of effecting the real changes in nature that posed a threat to flora and fauna. The consequences of large-scale land drainage and intensified cattle farming were noted. The lowering of lake levels, so essential to farming, was given special attention.

Gotska Sandön in the Baltic is one of the few national parks in the southern part of Sweden.

When the committee prepared its report, the lowering of the levels of two of Sweden's most bird-rich lakes then underway— Tåkern in Östergötland and Hornborgasjön in Västergötland— was noted disapprovingly. It was most unfortunate, the committee wrote, that such a project should be completed; projects involving the lowering of lake levels should be submitted to Academy scrutiny before being approved.

The report was sent to the Board of Crown Forests and Lands for comment. The board considered the grounds for the resolutions insufficient. None of the areas proposed as national parks had undergone scientific investigation, and the board itself lacked the expertise necessary to conduct such studies.

After the cool response of the board, the question of nature protection was shelved until the summer of 1907, when a new commission of inquiry was appointed within the Ministry of Agriculture. Karl Starbäck, Einar Lönnberg, and Justice Louis Améen were chosen to be the commission's experts. The latter brought not only necessary legal finesse to the commission, but as one of tourism's foremost spokesmen, he was already deeply engaged in the protection of nature.

The work of the commission resulted in two concrete bill proposals: a law on the protection of natural landmarks, and a law on national parks.[7] The former law was meant to make it possible for a landowner to protect natural landmarks on his property for—in principle—eternity. These landmarks could take the form of a remarkable rock formation, a swamp, the breeding grounds of a rare species of bird, or an "unusual stand of trees." The commission recommended a somewhat complicated procedure for establishing a nature preserve, in which the Academy would play a key role. A nature preserve could only be granted after an investigation and recommendation by the Academy, which would also draw up a national register of natural landmarks and take the initiative for new preserves. The commission also recommended that the Academy be given the exclusive right to request expropriation of areas to be set aside as natural landmarks, and that it should be assigned ultimate responsibility for natural landmarks on government lands.

In other countries—the example of Prussia being closest to hand—the responsibility for similar duties fell to a specific government institution. The commission, determined to hold down government costs as much as possible, felt that it was unnecessary to create a corresponding institution in Sweden. Instead, it relied on the good intentions of the Academy, and drew an analogy with Swedish legislation concerning "relics of the past." Just as the Royal Swedish Academy of Letters, History, and Antiquities was engaged in all-embracing work to preserve the relics of the past, "it seems only natural that the Royal Swedish Academy of Sciences devote its interest to the question of protecting the natural monuments of the nation." One might "view them as a sort of outdoor section of the Museum of Natural History, which is under the administration of the Academy of Sciences."[8] However, the analogy proved to be tenuous, as we shall see, and only made the Academy's job more difficult, and the efforts at protecting nature less effective.

The Academy was also granted a central role in the national parks. The commission's proposals were based on information collected earlier by the Committee for the Protection of Nature, along with proposals made by the Board of Crown Forests and Lands. The original idea, to designate as national parks areas that contained natural species and objects representative of each province, had been abandoned. The areas recommended by the commission all lay on crown lands, mostly in northern Sweden. Natural phenomena characteristic of Sweden's other regions were represented only sparingly. The explanation for this selectivity was that the commission was trying its best to avoid confrontation with economic interests. Hence, areas felt to be of little commercial value were preferred. Tourism interests, represented by Améen, had also played a decisive role, especially in the selection of national parks in the northern mountains.

It would be preferrable, according to the commission, that the administration of national parks be connected with the protection of natural landmarks, and that a thorough scientific study of the protected areas be implemented. Therefore, the administration ought naturally to be entrusted to the Academy of Sciences. It might seem like a major task—administering ten national parks, some of which covered enormous and isolated expanses. However, the commission was of the (somewhat naive) opinion that it ought not to prove such a heavy responsibility, since the goal of designating national parks in the first place was that the areas in question be left "to themselves." It was, however, understood that the Academy would incur certain expenses. To offset these, the Academy should receive an income from fines for violations of the nature protection laws, fees for entrance, camping, hunting, fishing, and hotel rights in the national parks, as well as sundry gifts and donations. Despite this, the commission added, the government would still need to provide the Academy with a small annual subsidy.

The commission's report was submitted to the Academy, which in turn appointed a committee composed of Gustaf Retzius, A.G. Nathorst, Gerhard Holm, and Hjalmar Théel. In its report, this committee stated that from a scientific standpoint, the nature protection proposal by the commission conformed to the opinions expressed by the Academy in 1905.[9] However, it was critical of one point: in contrast to the commission, the committee wanted the whole of Gotska Sandön protected and declared a national park. The committee also underlined the fact that the proposed manner of organizing protection would involve much more work for the Academy. It was not feasible to think that its members, already busy with diverse duties, would voluntarily and without remuneration take on this assignment as well. Nor could the Academy's own salaried employees be used. The only possible candidates for this job were the government-salaried intendents of the Museum of Natural History. But if they were to work on matters of nature protection, the museum would need more staff. Therefore, the Academy should accept this assignment only if all the departments of the Museum received new permanent, salaried assistants. Moreover, a bureau for the protection of nature should be created, employing someone to draw up and maintain the proposed national register of natural landmarks.

When the committee's report was discussed at the Academy at the beginning of 1908, it was toned down considerably after "lengthy discussion." The Academy approved in principle all of the commission's proposals for organizing the protection of nature, but avoided dictating the terms under which, according to its own members, it would be prepared to shoulder the new responsibilities. In hindsight this decision appears unfortunate, since the Academy thereby took on a major, pressing assignment without receiving sufficient funds for the work from the government.

Eventually, after minor legal adjustments, the proposals made by the Commission for the Protection of Nature formed the basis for two propositions in parliament regarding the natural landmark law and the national park law, both of which were passed in 1909. At the same time, further proof of the breakthrough of the nature protection idea was furnished when the Swedish Society for the Protection of Nature (*Svenska naturskyddsföreningen,* SNF), was founded, with the intention of "awakening and fostering love for our Swedish nature and working for its protection."

Nature's Monuments

The passing of these laws meant that the Academy of Sciences was entrusted with a key role in nature protection in Sweden. Its first major task was to fix more exactly the boundary lines of the national parks and to work out regulations and provisions for their administration. This work was carried out primarily by Einar Lönnberg, who now acted as the Academy's chief spokesman on nature protection. The committee appointed to hammer out national park regulations also included Christopher Aurivillius and Nils Gustaf Lagerheim. In time, as their duties increased, they were referred to as "members of the Committee for the Protection of Nature."

Those national parks falling under the administration of the Academy after the 1909 parliamentary decision included Abisko, Stora Sjöfallet, Peljekaise, and Sonfjället (all situated in the mountainous terrain of northern Sweden), along with Hamra, Garphyttan, Ängsö, and Gotska Sandön (although only a smaller portion of the island). Subsequently, other areas were added during the Academy's years as custodian of the nation's parks: Dalby-Söderskogen (1918), Vaddetjåkko (1920), Blå Jungfrun (1926), Norra Kvill (1927), Töfsingdalen (1930), and Muddus (1941).[10]

The bulk of the committee's work dealt with proclaiming various natural landmarks protected and issuing permits for activities within park boundaries, usually to conduct scientific research or to build hiking trails, sleeping accommodations, and other structures on behalf of the Touring Club.

One of the most time-consuming tasks involved selecting candidates for protection. This occasioned voluminous correspondence and entailed lengthy journeys when the inspection of a sight was called for. It would be impossible in the limited space available here to do justice to the amount of time and energy spent on such work; a good indication is provided by the lists of national parks and protected landmarks regularly published by the Academy, based on the national register it maintained. In the first

edition, published in 1919, protected objects numbered close to
100. Seven years later, there were 301. In 1932, 460 were listed; in
1938, 723; and in 1948 1,080.[11]

This list is further broken down according to the nature of
the particular objects, thereby giving a good picture of just what
sort of natural phenomena were thought worthy of saving for the
generations to come. Let us use the 1932 edition as an example.
Under the heading "Areas," we find 69 objects described vari-
ously as forest meadow, limestone formation, moraine area, bay
with islets, bird cliff, and so on. "Geological Landmarks" mentions 33 objects, the majority individual boulders. Under
"Plants" we discover 43 protected species found in one or more
of Sweden's counties, along with individual plants, flowers, and
stands. The latter grouping included by far the largest number of
protected objects, 371 out of a grand total of 460. This in turn
consisted mostly of single trees, usually oaks. Finally, under the
heading "Wildlife," 21 species that had been placed under protection in one or more counties are listed. In sum, as this compilation shows, it would not be misleading to claim that during
these years, the typical case before the committee concerned the
preservation of individual trees—"a large beech," "a great and
beautiful juniper," "a giant spruce," "a 'troll' pine," or, most
commonly, "a majestic old oak."

Organized protection of nature was, of course, a great step
forward. Sweden was now the first country in Europe with its own
national parks. And the ability to protect individual landmarks and
threatened species did indeed greatly advance the cause of nature
protection. Still, the actual direction and organization of the work
was far from satisfactory. For example, it seems that locations for
national parks were often hastily chosen. Once, plans to establish
a park on the Suorsa or Rissa crown lands in the forests of Tärendö
had to be abandoned—it was impossible to find a suitable location
there.

Moreover, resources available for the work were severely lim-
ited. This is perhaps best illustrated by the widely exploited anal-
ogy between legislation for the protection of nature and
legislation for the preservation of relics of the past, or antiquities.
While certain parallels did indeed exist between the Academy of
Letters, History, and Antiquities and the Swedish Antiquarian
Association on the one hand, and the Academy of Sciences and the
Society for the Protection of Nature on the other, in important

respects this analogy simply did not hold. Preserving relics of the past was not wholly dependent on voluntary efforts by members of the Academy of Letters or the Antiquarian Association; it was also supported by an august civil service department, the Central Office of National Antiquities. A similar institution was sorely lacking in the field of nature protection. In practice, three unsalaried members of the Academy's Committee for the Protection of Nature were doing the work that in its "analogous" case was being conducted by an entire civil service department. Obviously, the responsibility was a great one, and early work for the protection of nature was restricted by the tiny resources available to the Academy.

The analogy between relics of the past and natural landmarks can also be tested in another way. Early nature protection was "museum-like" in character. The concept "natural monument" [in Swedish, "naturminnesmärke"] has lost some of its original resonance today. So it is worth noting that the initial actions taken to protect nature involved precisely "natur*minnes*märken" ["minne" = "momento," "rememberance," "relic"]. Earlier, one had striven to preserve those relics of the past ["fornminne"] that bore witness to the history of man. Now, it was essential to preserve those monuments ["minnesmärken"] that spoke of the history of nature. To allow these living monuments to be destroyed and disappear would be tantamount to allowing Sweden's rune stones to crumble away, as someone said. Therefore, logic stated, natural landmarks ought to be seen as serving as the outdoor wing of the Museum of Natural History, or as a great national museum, an "artless Skansen,"[12] in the midst of nature.

This obsession with nature's "monuments" is explained by the fact that the scholars who were the driving force of the nature protection movement subscribed to the theory of evolution, which greatly informed the direction of their research. Quaternary geology and quaternary biology were "fashionable" disciplines. Varved clay, boulder ridges, bogs, and swamps were nature's own archives from which the history of the evolution of Sweden's flora and fauna since the last ice age could be traced. An idea of how the climate had changed and of how various species and plant communities had succeeded one another was being formed. The picture taking shape was one of nature in constant flux. This discovery of the history of the landscape recurs over and over again in the context of nature protection; one might even say that in essence the idea of protecting nature emerges directly from it. The aims of early nature protection were directed by interest in what was apprehended to be nature in its original state. Put simply, it was a matter of preserving for the generations ahead at least one boulder ridge, one primeval forest, one swamp, one waterfall, one forest meadow, one peat bog. Oddities and curiosities, known popularly as whims or freaks of nature, also generated excitement. The early literature of the movement is packed with such examples as curious rock formations, or a spruce with "branches which have grown into trees themselves." Much of the committee's protection activity also concerned this type of object.

In this context it is necessary to discuss the difference between conservation and preservation. Conservation is used mostly to refer to the idea of saving natural resources for future consumption. By preservation is meant "the attempt to maintain in their present condition such areas of the earth's surface as do not yet bear the obvious marks of man's handiwork and to protect from the risk of extinction those species of living beings which man has not yet destroyed."[13] Thus, preservation was aimed at *total* protection *against* mankind, while conservation aimed at protecting nature *for* mankind. Preservation and conservation can be seen as two ends of a continuum. At one end are the "non-use" preservationists, at the other the "wise-use" conservationists.[14] As may be understood from the examples provided above, in the beginning nature protection in Sweden was dominated by the idea of preservation.

Thus, early nature protection was not a matter of caring for and protecting all of nature, including those parts that had been tamed for the service of man. Protection was primarily afforded primeval, untouched nature and its monuments, oddities, and curiosities. As a result a paradoxical situation occurred in which a dynamic view of nature that stressed its evolution produced a mainly static preservation of nature. This conception of the aims of the protection of nature contained the seeds of many a conflict. For example, full-scale protection for one or another type of natural phenomenon did not guarantee protection from future change. As early as 1905, in a critical study of the Academy's first nature protection committee, botanist Gunnar Andersson indicated that total protection of forest meadows would lead to their being transformed into spruce woodlands.[15] It was later discovered that a number of the early protected areas—e.g., the national park Ängsö—were in fact examples not of untouched nature, but of highly cultivated land. The theory of protection therefore led to a result in direct opposition to its intent—the protected areas became overgrown and corrupted.

The overwhelming interest in primeval, untouched nature meant that environmentalist ideas (with the significance they have received in later decades) played a remote role in early nature protection work. Not that these ideas were irrelevant. For instance, at the beginning of the twentieth century, a debate raged over air and water pollution, with voices raised in favor of legislating against pulp mills and other polluting industries.[16] It passed by quite unnoticed in nature protection circles. Instead, the spokesmen of nature protection were careful not to oppose modernization and industrialization. In fact, they often spoke, in language colored by evolutionary rhetoric, of the struggle between nature and civilization, and saw as their proud duty the need to see the fight through to the finish—man's victory over nature. Only then, when the battle was won, would the victor show nature, the old enemy, the extent of his mercy.

It is not unfair to maintain that nature protection was subordinate to economic interests. The first major schism within the Swedish movement for the protection of nature also concerned the Academy's relationship with economic interests that exploited the forces of nature.

Stora Sjöfallet

In 1917, the State Power Board said that it wished to regulate the flow of the river Lule where it passed through Stora Sjöfallet national park. A dam located several kilometers above the majestic waterfall that lent the park its name would transform the river valley and large system of lakes running through the park into an artificial lake. This reservoir would be used by Porjus power plant and by a planned generating station at Harsprånget.

The Power Board conceded that regulation was at odds with the idealistic and scientific aims that had guided the decision to establish the national park. But as such great economic considerations were on the line, idealistic interests must yield to material ones. Therefore, the Power Board asked the Academy of Sciences to approve regulation of the river, which it did, although stressing that regulation would alter conditions in the park and reduce its value from a preservation standpoint. The interference was unfortunate, but considering "the enormous national economic benefit" connected with the project, the Academy felt that it could not reject the application. The only conditions posed by the Academy were that the encroachment be compensated for by enlarging another section of the park, and that regulation should proceed with the utmost attention to the natural environment.[17]

The decision to give up this section of the national park, which was made by parliament in 1919 after receiving the Academy of Science's approval, caused concern within nature protectionist circles. The strongest dissenting voice was that of the secretary of the SNF Thor Högdahl, who wrote in indignation that the mutilation of Stora Sjöfallet was a slap in the face of nature protection legislation. Among those sympathizing with Högdahl was Karl Starbäck. And yet, the movement was not unanimous in its condemnation. Högdahl was challenged by five members of the board of the SNF, who declared that the work of the nature protection movement would only suffer by clashing with economic interests. Among the five was Einar Lönnberg, who aside from leading the Academy's Committee for the Protection of Nature was also vice-chairman of the SNF.[18] It ended with Lönnberg and another member resigning from the SNF's board of directors.

In hindsight, it is easy to understand the indignation felt by the friends of nature. The Academy of Science's Committee for the Protection of Nature, the institution wielding the most power in questions pertaining to the protection of the natural environment, had let economic interests have their way. And in no less an instance than one concerning that "diamond in the crown of Swedish nature!"[19] Along with Gotska Sandön, Stora Sjöfallet was, from the moment the first suggestion about national parks came up, the most obvious example of an area of natural beauty that should be preserved for the generations to come. To this day, this classic example of the conflict between nature protection and economics stirs passionate emotions. Akkajaure—the artificial lake—stands as a monument to the fragility of nature protection legislation.

But the arguments of Lönnberg and the committee also deserve understanding. Their position was influenced by the charged atmosphere engendered by the ongoing world war. The regulation proposal surfaced when energy supplies were strained in Sweden. According to Lönnberg's calculations, if the same energy yield promised by regulation of the Lule were instead to be generated by steam, it would mean burning 300,000 tons of coal per year. For a nation hit by blockades and import problems during a world war, these figures spoke volumes. Moreover, hydroelectric power was invested with great expectations for rejuvenating the sparsely populated and neglected north of Sweden. Among other plans mentioned in the Power Board's application was that of building electrochemical plants in the north as part of the overall program for developing hydroelectric power. Many would have considered opposition to such development as irresponsible. The area, according to Lönnberg, was in no way unique. It would certainly be possible for preservationists to find endless expanses of land in Lapland to compensate for the loss.

Some compensation was made as early as 1918–1919, when two new national parks were established—Vaddetjåkko in the far north and Dalby in the south. The origins of the idea for Vaddetjåkko can be found in Lönnberg's proposal that a park be set up north of Torneträsk. The Academy of Sciences and its committee also played an active role by urging the establishment of Dalby (thereby creating the first national park in Scania and southern Sweden). It should also be noted that this particular national park was opened in direct *contradiction* to economic interests. The Board of Crown Forests and Lands opined that the woods of Dalby were of great economic value; yet, as we read in the government proposition, "in this case, economic interests must bow to historical and scientific ones."[20]

However, the gift of these new national parks can still not disguise the fact that in the eyes of many friends of nature, the Committee for the Protection of Nature had been weak and compliant. Nor was this conclusion tempered by the fact that the planned hydroelectric station, the original motivation for placing a dam in the park, was not being built. Declining coal prices and difficulties in finding a market for electricity forced the Power Board to stop work on Harsprånget Power Station in 1922 (it was not resumed until 1945). At that point it might still have been possible to save the park; however, the Academy refrained from raising the matter anew, and the dam was completed.[21]

The Need for a Government Authority

For those critical of the Academy's position, the measures taken in Stora Sjöfallet national park again brought to the fore the question of the organization of nature protection in Sweden. In

1924, the board of the SNF took up the call for a government authority to deal exclusively with nature protection. In an appeal to the minister of agriculture, the chairman of the SNF, Rutger Sernander, enumerated one instance after another in which conflict had arisen between economic and protectionist interests. Due to the lack of orientation and planning in nature protection work, the interests of protection suffered. According to Sernander, the Academy of Sciences lacked the necessary resources for the demanding work involved in the protection of nature. Its appointment as expert and administrative authority was meant to be provisional. Therefore, it was high time to create a new civil service department for the protection of nature.

In its reply, the Academy defended itself against these charges. It denied that its appointment was merely temporary, and insisted that both the existing nature protection legislation and the resources at its disposal were sufficient. The Academy also used this opportunity to criticize the SNF. Instead of promoting the development of nature protection, the SNF had publicly ridiculed those measures taken by the Academy. Nor did the Board of Crown Forests and Lands, in its response to the SNF's appeal, feel that there was any need for a new civil service department. The Academy numbered among its members representatives for all the branches of natural science relevant to the field of nature protection, and was therefore competent to pass judgment on any matter that might occur with the greatest skill.[22]

However, the SNF's criticism was not ignored by the Academy. It took measures to strengthen its leadership of the nature protection movement, and in early 1925 new regulations for dealing with matters of nature protection were adopted. The Committee for the Protection of Nature's position was consolidated, and its membership increased to five. To guarantee the necessary competence, it was agreed that two of the committee's members must be experts in botany, two in zoology, and one in geology. The committee was also strengthened by being granted the power to make all decisions. Only in instances when the committee was unable to reach unanimous consent and any of its members requested it would a question be referred to the Academy itself for a final decision.

The changes paid off. That much can be ascertained by perusing the Academy's publication series in matters pertaining to the protection of nature. Before 1925 (i.e., during its first 15 years), only two publications in the series had come out. In the following 15

years, however, nearly 40 works were published in the same series. Most were descriptions of the flora and fauna found in the
national parks administered by the Academy of Sciences.

However, the divisions between the Committee for the Protection of Nature and the Society for the Protection of Nature
remained. This is explained mainly by the divergent concepts held
by the two groups about the goal of the protection of nature.
Leading representatives of the SNF (primarily Högdahl and Sernander) advocated a "preservationist" stance, while the committee began gravitating towards a more "conservationist" outlook.
A major bone of contention was the development of tourism in the
national parks. The Academy was generous in granting permits to
build hiking trails and lodges and other structures within the confines of the parks. This was roundly condemned by Sernander and
others, who viewed *any* encroachment on nature as evil. The
Academy explained that it repudiated the idea that the most attractive parts of the mountain landscapes were to be made attaina-
ble only to those "who were fortunate enough to have both the
time and the wherewithall to embark upon formal expeditions."
To reserve "the best of mountain nature for a handful of the
chosen seems to the Academy to have little to do with the protection of nature as the Academy defines this term."[23]

During the 1930s, the antagonism between the Academy and
the SNF lessened as the "conservationists" gained the upper hand,
even within the ranks of the SNF.[24] However, Rutger Sernander,
who had resigned his post as chairman of the society in 1930,
continued his drive to replace the committee with a civil service
department. In 1933 he was appointed by Arthur Engberg, the
minister of education and ecclesiastical affairs, to study the question of whether the handling of nature protection matters was
becoming increasingly routine and standardized. This resulted in
an official report in 1935, in which Sernander again proposed the
establishment of a civil service department for the protection of
nature with far-reaching authority.[25]

In its response to this recommendation, the committee agreed
with the need for a structural reorganization. The matters it had
to deal with had grown not only in number but in complexity,
which was why it was unreasonable to expect the committee's
members to devote as much attention to them as they rightly
deserved. Still, it rejected the idea of a new civil service department. The protection of nature must remain the domain of the

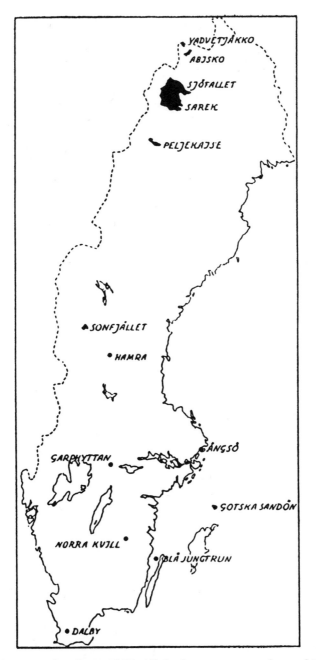

Sweden's national parks in 1927. All the larger ones are located in Norrland, the northern part of Sweden. From C.J. Anrick, *Våra svenska nationalparker* (1927).

natural sciences. Instead, one ought to build on the existing foundation and strive to preserve the connection with the Academy of Sciences. To realize this goal, the committee exhumed an old idea: the protection of nature ought to be joined to an institution administered by the Academy—the Museum of Natural History. An institute for the protection of nature should be established there.[26]

Neither Sernander's nor the committee's recommendations were ever realized. One of the reasons they were not was that the definition of "the protection of nature" was being expanded to embrace "environmental protection" during the 1930s. This meant that an eventual reorganization would be coupled to other matters still under debate. The protection of nature was no longer seen as the exclusive property of the natural sciences. Scientifically motivated protection came to be complemented by a socially motivated protection of nature, as other, more activist environmentalist ideas began to be heard. This was already reflected in the committee's response to Sernander's study. There, it was stated that the protection of nature implied more than just the ambition to preserve untouched nature; it should also "see that the encroachments made upon nature necessary to cultivation are conducted with respect, avoiding needless damage to the beauty of the landscape, its natural treasures and its plant and animal life." In 1937 a parliamentary proposal expressed the wish that a new law for the protection of nature be worked out. There, for the first time, current environmental problems were coupled with the orientation of the protection of nature. It was stated that the most important question in the protection of nature was "how to properly counteract the polluting of our lakes and rivers, which as a result of continuing settlement and industrialization, assumes more and more alarming proportions."[27]

The motion was defeated, but not because parliament did not agree with its essence. Instead, it referred to other on-going studies, which it was presumed would take into account questions pertaining to nature protection. Reference here is made to the so-called "leisure inquiry," which examined ways of making it easier for the urban populace to enjoy time out in the open air and in the midst of nature, and to an inquiry into legislation regarding measures to halt air and water pollution.[28]

The advent of the Second World War caused the much discussed reorganization of nature protection to be postponed. However, the question continued to be raised, especially by the

Academy of Sciences, which repeatedly called attention to it in its supply estimates. In 1946 the Academy requested, at the behest of the Committee for the Protection of Nature, that the government hold a new inquiry into the organization and administration of nature protection. The duties, it insisted, had grown to an extent unimaginable 40 years earlier, the reason "being the changes which to a greater and greater degree affect and continue to affect the Swedish landscape in connection with the progress of material society." Protection now had to be extended to encompass features of the old cultural landscape, since modern agricultural methods, increased settlement, the extension of the communications network, and the exploitation of rivers and other natural resources had altered the very appearance of the countryside. Furthermore, nature protection had gained a social dimension, since sporting and leisure interests now also had to be considered.[29]

The Academy's request resulted in the government establishing an "inquiry into the protection of nature" in 1946, which four years later recommended a new nature protection act.[30] This inquiry laid the groundwork for a new law, passed in parliament 1952, and for a proposal to set up a state-controlled central office for the protection of nature. That proposal was defeated again during the debate on how to organize nature protection. Still, the new nature protection law represented significant changes as far as the duties of the Academy were concerned. Administrative responsibility for the national parks was transferred to the Board of Crown Forests and Lands, along with the administration of natural landmarks on crown lands and the task of keeping a register of the country's natural monuments. The Academy's duties were henceforth to be of a more investigative and consultative nature, concerning the scientific aspects of nature protection. Thus, the Academy was placed on equal footing with two other voluntary organizations: the SNF, which would provide a general pool of expertise, and the Society for the Preservation of Local Monuments, which would attend to landscape care on a more local level.

The Water Regulations

The new law emphasized the Academy of Science's standing as the body to which all matters concerning the protection of nature must be referred when scientific interests were involved. Since scientific opinion was relevant in nearly all matters of nature protection, the Academy continued to have plenty of work. Indeed, although the Board of Crown Forests and Lands had assumed many of its previous responsibilities, the Committee for the Protection of Nature found its workload increased during the 1950s.

Above all, beginning in the 1940s and increasing from the mid-fifties to the beginning of the sixties, one issue dominated: water power. The rapid development of hydroelectric power in Sweden begun during the Second World War resulted in many confrontations with nature. Thus the committee's heaviest responsibility, as it saw it, became "protecting anything of scientific value which might be threatened by the ever-increasing exploitation of our water power resources."[31] In time, a sort of routine was developed wherein the committee conducted its investigations in two stages. When a project to regulate a water source was proposed, the Judicial Board for Public Lands and Funds and the Water Rights Court assigned the committee the task of organizing and executing field surveys of the areas in question to determine whether anything of scientific value was endangered. Only rarely did these broad surveys result in an area being spared from exploitation. In the second stage, which was reached after the decision to build was made and was more often than not financed by the power interests, the committee organized a documentary study of the plant and animal life in the areas affected by the construction project.

This manner of working, to try and accumulate knowledge about the flora and fauna threatened by economic exploitation, had a long tradition at the Academy. As early as 1908, the Academy had received its first such commission, which concerned the proposed drainage of the unique lake Tåkern. Parliament had approved the project, since it was considered invaluable for local agriculture. However, it had also allotted funds to the Academy, which requested permission to undertake a comprehensive biological survey before the draining was completed.[32] The encroachment on Stora Sjöfallet resulted in a similar survey when the government, as a condition before regulation could start, demanded scientific studies of the areas destined to end up underwater.[33]

The water regulation questions dealt with by the committee throughout the 1950s and early 1960s were extensive, and made great demands on its members. A compilation of the regulation projects upon which the Committee commented between 1957 and 1961 covers no fewer than 57 cases. In many instances this led to costly, large-scale documentary studies.[34] The result was a great number of reports, inquiries, and dissertations that, taken together, comprise a massive inventory of the natural life of the lake systems and river valleys transformed forever by the rapid development of hydroelectric power in Sweden.

The development of hydroelectric power, which seemed poised to change irrevocably the face of the natural landscape in the areas involved, especially in northern Sweden, was naturally controversial. During the 1950s the exploitation of lakes and rivers monopolized the environmental question and engaged the public to an unprecedented degree. Debate in newspapers and periodicals was heated, and a chorus of nature, tourist, and outdoor life groups took up the cry to save imperiled nature.[35] Nor was the Committee for the Protection of Nature satisfied with simply passively documenting the areas destined for exploitation. It participated much more actively than ever before in marshalling opinion against the advocates of water power. One of the more noted initiatives taken by the committee (staged to rally support against planned construction in Torneträsk and the Torne and Kalix Rivers) was a circular sent to approximately 200 Swedish and foreign scientists, who were asked their opinions on the proposed project. Excerpts from the statements (100 of which came from abroad) were selected and published in book form and presented to the government.[36]

In 1954, the committee together with other organizations assembled a list of rivers it felt should be exempted from future exploitation. From that project was born the so-called "Environmental Protection Delegation," initially consisting of two members appointed by the government plus representatives of the SNF and the Society for the Preservation of Local Monuments. Soon, the chairman of the Committee for the Protection of Nature was also involved in the Environmental Protection Delegation.

The delegation's main task was to reach agreements with the representatives of hydroelectric power stipulating which lakes and river stretches were to be spared. The most famous result was the "Peace of Sarek," 1961, in which the delegation and the Power Board signed an agreement to save 28 cases. In return, the representatives of environmental protection approved of a similar number of cases which the board would be allowed to exploit. Thanks to this agreement, which was considered a great victory for the cause of environmental protection, Sarek and Padjelanta national parks were spared serious encroachment.

Despite the relative success of the delegation, it became apparent during the 1950s that environmental protection legislation and the organization of nature protection were still not what they should be. It was not just the sometimes uneven battle against the powerful hydroelectric interests that made the serious breaches in the splintered organization apparent. Rapid economic growth after the Second World War brought with it a massive intensification of the exploitation of natural resources. This exploitation of the country's natural assets had, as was verified by two parliamentary motions in 1959, been influenced or modified only slightly by the law for the protection of nature. The care and protection of Swedish nature was at a critical stage, according to the sponsors of the motion. It was therefore necessary once again to rework the law and establish "a government commission for the care and protection of the environment."[37]

The motions resulted in yet another inquiry into the organization of nature protection, and in 1962 the establishment of the National Environmental Protection Commission was proposed. The proposal was passed by parliament, and the new commission began its work the following year. Awareness was also growing of the intimate connection between environmental protection and questions normally dealt with by other government authorities. Recall that it was during the 1960s that terms such as "environment," "ecology," and "environmental control" became common currency in public debate. As far back as the 1950s, a state water conservation authority, the Water Inspection Board, had been established. In 1964, a state air pollution control commission came into being; and in 1967 the various commissions and boards were brought together in a new civil service department, the National Environmental Protection Board.[38]

A New Epoch

The great organizational changes made during the 1960s spelled the end of the Academy of Sciences and its Committee for the Protection of Nature as official environmental protection agencies. Thus, a 50-year tradition was laid to rest. Yet it was also the beginning of a new epoch in the Academy's engagement in questions concerning the protection of the environment.[39]

Awareness of the complexity, urgency, and global proportions of environmental matters, which slowly but surely grew during the 1950s and was subsequently firmly established in the 1960s and 1970s, led to an intense focus on the need for increased scientific contributions. The past two decades are still too recent to be placed in historical perspective. However, there is no doubt that the scientific activity that developed as a result of global environmental change (and which, of course, is not just characteristic of the Academy of Sciences) will prove to be one of the most significant elements in the history of science in the twentieth century. Great indeed is the contrast between the relatively narrow preservationist interests that continued to influence scientific nature protection at the onset of the 1950s and the basic research carried out as a result of the environmental protection problems that became so obvious just ten years later.

It was no longer a matter of preserving individual objects and areas in nature for the needs of science. Instead, it was necessary to acquire knowledge and insight to deal with environmental pollution and the waste of resources endangering the very life of our planet. And this role suited the Academy of Sciences much better than its earlier one as official environmental protection agency.

It is hardly an exaggeration to maintain that global environmental and energy problems determined the orientation of the work of the Academy throughout the 1970s. The Academy form proved to be invaluable when it was a question of drawing attention to and discussing difficult environmental problems. The Academy became a forum for debate, conferences, symposia, and other meetings, in which a variety of environmental and natural resource questions were delved into and the Academy's scientific versatility, its position as a free and independent institution, and its rich network of international contacts were finally given their due. It would be impossible in the limited space remaining even to hint at the richness of the variegated questions and initiatives taken up under the auspices of the Academy in the past two decades; a few examples will have to suffice: the symposium "Energy in Society," 1973; the international Nobel symposium held in 1976 on "Nitrogen—Essential Factor of Life and Growing Threat to the Environment"; and a conference on phenoxyacetic acids and dioxins in 1977.

Research at the Academy's older institutes, e.g., the Kristineberg Marine Biological Station, also became more relevant when worries began to be expressed over the condition of the soil, water, and air. Furthermore, the Academy has opened a new institute; thanks to a donation from the Kjell and Märta Beijer Foundation, the Academy established an international institute for energy and human ecology in 1977.

Another indication of the Academy's deep engagement in environmental matters is *Ambio,* a monthly periodical launched by the Academy in 1972 in connection with the United Nations Conference on the Human Environment in Stockholm. *Ambio* has become an internationally renowned environmental journal, now published in co-operation with the World Resources Institute in Washington and others.

The old Committee for the Protection of Nature had played out its role. In 1974 it was replaced by an Environmental Protection Committee with the task of monitoring from a scientific perspective nature and environmental protection in Sweden and the world on the Academy's behalf. The new committee also functions as the Swedish branch of SCOPE (Scientific Committee on Problems of the Environment). Chosen as the committee's first chairman was a "layman," county governor Rolf Edberg, one of Sweden's most recognized writers on environmentalism.

In 1987 the Academy's engagement in environmental protection was overhauled, resulting in the establishment a year later of a special secretariat for the environment, with a permanent secretary of environmental affairs at the Academy. The Environmental Protection Committee was strengthened in other respects as well, and its name has now been simplified to "Environmental Committee." Finally, it should be noted that the Academy is host for the Secretariat of the "International Geosphere-Biosphere Program—A Study of Global Change," which is the coordinating body for the environmental program of the International Council of Scientific Unions (ICSU).

Notes

1. Report of the Secretary in *KVAÅ* 1910. These yearbook reports, often supplemented by a special section dedicated to the work of the Com-

mittee for the Protection of Nature at the Academy, provide the main source of information in this account for the years 1910–1968. The activities of the Academy in the field of nature protection can also be followed in the reports presented in the yearbook of the Society for the Protection of Nature (SNF), *Sveriges natur,* from 1913 onward. References drawn from these sources will generally not be provided with a footnote.

2. A.E. Nordenskiöld's article is reprinted in *Sveriges natur,* 1911.

3. The early history of nature protection in Sweden has been treated in detail by Désirée Haraldsson, *Skydda vår natur! Svenska Natur-skyddsföreningens framväxt och tidiga utveckling,* Bibliotheca Historica Lundensis 63 (Lund, 1987). [Summary in English].

4. *Betänkande rörande åtgärder till skydd för vårt lands natur och naturminnesmärken. Afg. af inom K. Jordbruks-departementet för ändamålet tillkallade sakkunnige* (Stockholm, 1907), 6 ff.

5. Parliamentary publications, Lower House motions 1904:194.

6. *Yttrande öfver riksdagens skrivelse ang. naturskydd* (Stockholm,1905).

7. *Betänkande rörande åtgärder till skydd för vårt lands natur och naturminnesmärken. Afg. af inom K. Jordbruksdepartementet för ändamålet tillkallade sakkunnige* (Stockholm, 1907).

8. *Ibid.,* 30 f.

9. *Yttrande öfver betänkande rör. naturskydd* (Stockholm, 1908).

10. Bo Rosén, "Nationalparkerna," in *Fridlyst* (Stockholm, 1960).

11. Aurivillius, Chr., *Förteckning å svenska nationalparker samt å natur-minnesmärken, som blivit fridlysta under åren 1910–1918* (K. Svenska Vetenskapsakademiens skrifter i naturskyddsärenden nr 1, Stockholm, 1919). Four more editions were published under the title *Förteckning å svenska nationalparker samt å fridlysta naturminnesmärken.* Henceforth they are referred to simply by the compiler's name, its number in the Academy's series on matters pertaining to the protection of nature, and the year of publication: Aurivillius, Chr. (no 6, 1926); Almqvist, Erik and Florin, Rudolf (no 21, 1932); Florin, Rudolf (no 34, 1938); and Florin, Rudolf (no 47, 1948). The latter two editions also contain detailed bibliographies of the literature available on national parks, natural landmarks, and other nature protection matters.

12. Skansen is the name of the large open-air museum built at the end of the nineteenth century in Stockholm, displaying examples of the old peasant culture of Sweden.

13. John Passmore, *Man's Responsibility for Nature* (New York, 1974), 73.

14. Haraldsson, 213.

15. Gunnar Andersson, "I Sverige under senaste tid företagna åtgärder till naturens skydd," in *Ymer,* 1905.

16. See Lars Lundgren, *Vattenförorening: Debatten i Sverige 1890–1921,*

Bibliotheca Historica Lundensis 30 (Lund, 1974). [Summary in English].

17. Parliamentary publications, extraordinary session of parliament 1919, prop. no. 3.

18. Thor Högdahl, "Stympandet av Stora Sjöfallets nationalpark," in *Sveriges natur* 1920, 140; Starbäck and Lönnberg in *Tidskrift för hembygdsvård* 1920, 143 f. and 174 ff., and *Sveriges natur* 1920, 171, respectively.

19. This epithet is found in Sigfrid Ericson, "Skövlingen av vår främsta nationalpark," in *Sveriges natur* 1921, 16.

20. Parliamentary publications, prop. 1918:9.

21. Sigfrid Ericson, "Skövlingen av vår främsta nationalpark," in *Sveriges natur* 1921, and *Sveriges natur* 1922, 147 ff.

22. Sernander's appeal and the response of the Academy are reprinted in *Betänkande med förslag rörande det svenska naturskyddets organisation och statliga förvaltning* (Governmental reports, SOU 1935:26), 153 ff.

23. *Sveriges natur* 1927, 160.

24. See Haraldsson, 202 ff.

25. *Betänkande med förslag rörande det svenska naturskyddets organisation och statliga förvaltning* (Governmental reports, SOU 1935:26).

26. K. *Vetenskapsakademiens naturskyddskommittés förslag till yttrande över Sernander: Betänkande ang. svenska naturskyddet* (Stockholm, 1936). See also *Betänkande* 1936, 185 ff.

27. Parliamentary publications, Upper House motions 1937:40, and Lower House 1937:99.

28. Parliamentary publications, first report of the standing committee on civil-law legislation, 1937:56.

29. Communication reprinted from *Förslag till naturskyddslag m.m. Betänkande avgivet av Naturskyddsutredningen* (SOU 1951:5), 57 f.

30. *Ibid.*

31. Quoted in "Kungl. vetenskapsakademiens naturskyddskommitté, dess tillkomst och arbetsuppgifter," unsigned memorandum from 1962 in the archives of the Committee for the Protection of Nature at the Academy of Sciences.

32. Proposition 1908:215. In the end, the lake never was drained, for economic reasons. However, the scientific survey done at Tåkern *was* completed, and resulted in a series of essays published by the Academy of Sciences as *Sjön Tåkerns fauna och flora* (1915–21).

33. Among others, these studies resulted in Gunnar Björkman, *Kärlväxtfloran inom Stora Sjöfallets nationalpark jämte angränsande delar av norra Lule lappmark,* published in 1939 as the second number in the Academy of Sciences' series of dissertations on nature protection questions.

34. See the compilation in "Kungl. vetenskapsakademiens naturskydds-kommitté, dess tillkomst och arbetsuppgifter."

35. Bo Rosén, *Den glömda miljödebatten* (Stockholm, 1987); and Evert Vedung, "Striden om de strömmande vattnen", in *Daedalus 1984*. [Summary in English].

36. *Bör Torneträsk regleras? Uttalanden av in- och utländska vetenskapsmän med anledning av planerna att för vattenkrafts-exploatering bygga ut Torneträsk samt Torne och Kalix älvar med tillflöden* (Kungl. svenska vetenskapsakademien, Skrifter i naturskyddsärenden nr 50, Stockholm, 1960).

37. Parliamentary publications, Upper House motions 1959:43, and Lower House 1959:51. See also third report of the standing committee on civil-law legislation 1959:23.

38. The growth of the State Environmental Protection Board is treated in Lennart Lundqvist, *Miljövårdsförvaltning och politisk struktur* (Stockholm, 1971). [Summary in English].

39. The following is largely based on the yearly reports by the Academy of Sciences, published in the Academy's series *Documenta* 1974–1988; "Kungl. Vetenskapsakademiens engagemang i miljöfrågor. Rapport från Kommittén för översyn av KVA:s miljövårdsverksamhet" (communication to the board of the Academy 1987–08–17); and, Carl Gustaf Bernhard, "Om Vetenskapsakademiens engagemang i aktuella miljö, naturresurs- och energifrågor 1972–1980" (manuscript, September 1987).

ELISABETH CRAWFORD

The Benefits
of the Nobel Prizes

To bestow honors is a tradition in academies and learned societies; it is also one of their *raisons d'être*. Once these bodies had been constituted, election to them became an honor in its own right and remained so. Shortly after their inception, the Royal Society of London and the French Academy of Sciences received donations that permitted them to institute prizes and medals, respectively, in 1709 and 1715. The new-fledged Royal Swedish Academy of Sciences followed suit in 1754 when it established its first prize. Thus, by the eighteenth century, the stage was already set for the long celebration of scientific achievements through prizes and medals that was to take on a new dimension on the eve of the twentieth century with the creation of the Nobel prizes.[1]

The Royal Academy of Sciences has played a pivotal role in the creation and continued functioning of the Nobel institution in two important respects. First, it was entrusted at the outset with two prizes, in the neighboring disciplines of physics and chemistry. This gave the viewpoints and actions of the Academy particular weight in formulating the rules and procedures for awarding the prizes. The part played by the Academy in the Nobel institution was further enhanced by the creation of the prize in economic sciences in memory of Alfred Nobel in 1968. Second, the two initial prizes in the Academy carried particular weight in monetary terms. My focus in this article is not the award decisions, but the use of the money that was set aside for expenditure in connection with the awards. With this money the Academy could equip and maintain research laboratories in physics and chemistry, the

Nobel institutes. The discussions and decisions concerning these ran parallel to and occasionally intersected with the actual award decisions during most of the period up to World War II. The Nobel institutes fitted a general trend early in the century towards setting up institutes and laboratories for scientific research outside the university; some were independent, others attached to a foundation or academy. In this respect, they are part of the history of scientific organization during the first half of the twentieth century.

The Genesis and Concretization of the Prizes in the Academy

Alfred Nobel died on December 10, 1896, leaving, apart from his enormous fortune, a will that prescribed in rather general words that his assets be used to promote science and culture. The most important scientific discoveries in the fields of chemistry, physics, and medicine were to be rewarded; furthermore, a peace prize and a literature prize were to be established. In Nobel's not quite crystalline words, the prizes should go to "those who, during the preceding year, shall have conferred the greatest benefit on mankind."

Before any prizes could be awarded, the will had to be translated into directives that were legally and practically manageable. This process turned out to be extraordinarily complicated and full of bizarre highlights. Nobel's family initially protested against the entire project and pursued the line that the will was legally invalid. The sensitive, diplomatically demanding assignment of putting Nobel's will into practice was carried out by the two executors, Ragnar Sohlman—who told the story in his book *The Legacy of Alfred Nobel*[2]—and Rudolf Lilljequist, both engineers. The most important and extensive work, however, was done by the executors' legal representative, the independent and tireless Carl Lindhagen, later mayor of Stockholm. After a comet-like career in the courts, Lindhagen had entered politics and was elected to the lower house of Parliament for the Liberal Party in 1897, the same year as he was engaged by the executors.

It was Lindhagen and Sohlman who negotiated the financial settlement that ended the opposition of the Swedish relatives of Nobel to the will. Under the agreement reached in 1898, the Swedish relatives received a few million kronor to share in addition to the legacies they had been given in the will. By then the Russian branch of the family, headed by Nobel's nephew Emanuel, had relinquished their share of the estate in return for the controlling interest in the Nobel Brothers Naphtha Company in Baku. The settlement meant that Nobel's estate of about 30 million kronor, less the 10 percent paid to the relatives, could be invested as he had prescribed "in safe securities," and used for the purposes for which it had been intended. When first awarded in 1901, each of the five Nobel prizes amounted to about 150,000 kronor, a sum that surpassed all previous awards of this kind.[3]

Under Nobel's last will and testament, the Academy of Science was to reward the "person who shall have made the most important discovery or invention within the field of physics," and the person who had made "the most important chemical discovery or improvement." To do this, the Academy was to receive two-fifths of the interest on the Nobel fund. The provisions of this will could have caused keen disappointment had Nobel's earlier will dated 1893 been more widely known. In this document, the Academy of Sciences was the sole institution designated to award an unspecified number of prizes "for the most important and pioneering works within the wide domain of knowledge and progress." It was to have received 64 percent of Nobel's fortune; the remainder would have constituted bequests shared between Nobel's family and several Swedish institutions, among them the Karolinska Institute and the Stockholm Högskola.

Alfred Nobel had been a foreign member of the Academy belonging to the economic science section since 1884. He had let his sponsor Adolf Erik Nordenskiöld know at the time of his election that he regarded it "not as a reward for the little I have been able to accomplish but as an encouragement for future work."[4] As a foreign member of the Academy, Nobel must have realized that his bequest represented an *embarras de richesse* in relation to existing resources. Although the Academy's Proceedings and journals were major platforms for the publication of research by Swedish scientists, the prizes and grants of the Academy were neither numerous nor lavish. The largest grant by far was the Letterstedt travel stipend (4,000 kronor). Nobel's bequest was not primarily intended to aid Swedish science and scientific institutions, however, but was meant to further progress in science and culture internationally, as indicated by the passage in the will that "the most worthy shall receive the prize, whether he be a Scandinavian or not."

From the start of the negotiations between the executors and the future prize awarders, the latter let it be known that they expected to receive some compensation for carrying out the onerous task of selecting prizewinners.[5] There was a tradition of paying stipends or *épices* to the juries of prize competitions that dated back to the first competition launched by the French Academy of Sciences in 1719. The size of the Nobel legacy, however, made the future Swedish prize awarders more ambitious. They formulated their demands in reply to the executors' letter of March 24, 1897 in which each institution was asked if it was willing to accept the assignment of Nobel's will. In the Academy, the reply was prepared by a five-member committee, including among others, the Academy physicist, Bernhard Hasselberg, and its secretary, Georg Lindhagen, Carl's uncle. The committee recommended first that the two portions of the fund allowed for the prizes in the Academy should include money to meet expenses, mainly honoraria paid to members of the prize juries. Second, it suggested that prizes that had not been awarded due to a lack of worthy discoveries and inventions remain with the Academy and the money be used for the same purposes as the Nobel prizes were intended to serve. In this connection, the idea was introduced of a "well-equipped Nobel Institute for Physics and Chemistry to which foreigners, too, should have access in order to carry out scientific investigations."

With the Academy's reply in preparation and positive replies already at hand from the Norwegian Storting, the Karolinska Institute, and the Swedish Academy (of literature and language), the executors and their attorney clearly felt that official negotiations over the statutes of the Nobel foundation could begin. This feeling was bolstered by the decision of the attorney general to consider Nobel's donation a matter of national interest to Sweden, which made the government a party to any legal action that had to be taken in defence of the will. Their optimism melted into thin air, however, when on June 9, 1897, the Academy of Sciences meeting in plenary session decided to reject the recommendation of its committee, and not accept the mandate to award Nobel prizes. This about-turn came through the persuasive powers of one member, Hans Forssell, an economic historian and former member of

Alfred Nobel (1833–1896), the famous benefactor, photographed during
the last year of his life. (Courtesy of the Nobel Foundation.)

the government. Shortly before, he had failed to dissuade the Swedish Academy, of which he was also a member, from accepting responsibility for the literature prize. He now convinced the members of the Academy of Sciences that if it were to discuss its role as a prize awarder before the will had been probated, it would only encourage the Nobel relatives to start a court action. As a result of the change of heart in the Academy of Sciences, official negotiations with the future prize awarders could not start until late in 1898 when the settlement with the Swedish relatives mentioned earlier had been accepted by all the prize-awarding institutions as well as by the Swedish government.

Rather than impair the outcome of the official negotiations, however, the Academy's refusal probably enhanced it. First, it made for a pause that gave the different interested parties time to reflect on the many suggestions advanced in 1897 concerning procedures for selecting the prize winners. More important, the Academy's refusal even to name delegates to negotiate with Sohlman and Lindhagen did not prevent some of its members from becoming involved unofficially. This was to be of the utmost consequence in the final outcome, since many of the proposals advanced by this group were more innovative than the recommendations made by the committee charged with formulating the reply of the Academy to the executors' letter of March 24, 1897. During this creative phase, which lasted for about a year, alternative arrangements for how the Nobel fund should be set up and the prizes awarded were discussed more openly and freely than would have been possible in official negotiations with representatives of the prize-awarding institutions.

The most innovative proposals originated in a small group centered on the Academy of Sciences but with strong ties with Stockholm Högskola, a private non-degree-granting institution primarily oriented toward the sciences, which had opened in 1878. The group included: Lindhagen, the attorney to the Nobel estate, who had served as Secretary of the Högskola (1884–1894); Lars Fredrik Nilson, who had chaired the committee that had recommended that the Academy accept the task of awarding the prizes; Otto Pettersson, a member of the Academy and professor of chemistry at the Högskola; and Svante Arrhenius, professor of physics at the Högskola, whose international reputation in physical chemistry was well established, but who was not yet a member of the Academy.

The members of this group all had a strong interest in the creation of a Nobel institute for scientific research linked with the Högskola. From its original nebulous state, the idea of such an institute took concrete form in informal discussions with the executors early in 1898. The proposal that emerged was for a large Nobel institute with laboratories for experimental work in physics, chemistry, and physiology and a common library all forming a single unit. This was a novelty, for while academies had long derived benefits from awarding prizes, these benefits had not hitherto involved the creation of their own research institutes.[6] Nevertheless, the Academy had a tradition of maintaining its own research units, although the chemistry institute had fallen in to abeyance by the mid-nineteenth century, and the physics one, headed by Academy physicist Bernhard Hasselberg, only functioned on a modest scale.

In the proposal, the bond with the prize system was maintained through the dual rationale that was presented for the Nobel institute: on the one hand, it would function alongside the prize juries and carry out "detailed scientific investigations to determine the factual basis of a discovery and its significance"; on the other, it would help Swedish science acquit itself honorably of the task of awarding the prizes by creating a "critical mass" of personnel and facilities in physics and chemistry. In Arrhenius's and Pettersson's vision, the Nobel institute, located on Observatory Hill with the Högskola as its closest neighbor, would provide the capital with a science center that would enable it to outshine the provincial university towns of Lund and Uppsala.

The document proposing the Nobel institute contained another innovation: the idea of soliciting nominations for the prizes from outside experts. This idea was put forward by Arrhenius at one of the unofficial negotiating sessions early in 1898. His point of departure was the principle, affirmed in earlier memoranda, that the selection of prize winners should not be based on applications. He suggested instead that certain categories of scientists should have the right to nominate candidates. For prizes awarded by the Academy of Sciences, this right would be vested in Swedish and foreign members of the Academy and could also be extended to the staff of the science faculties at Uppsala, Lund, and Stockholm universities as well as to Nobel laureates.

These proposals were incorporated into the draft that Lindhagen presented as "a basis for discussion" when official negotiations over the statutes of the Nobel Foundation finally began early in 1899. In these, the interests of each future prize awarder, represented by its official delegates, were the dominant element. The delegates from the Academy of Sciences—L.F. Nilson and Bernhard Hasselberg—sought above all to maintain the original idea of the large Nobel institute in the statutes. They had to contend, however, as did Lindhagen, with strong opposition from the delegates of the Karolinska Institute led by its rector Karl Mörner. In the broadside that the Karolinska delegates launched against Lindhagen's draft statutes, their rejection of a Nobel institute common to the Academy and the Karolinska Institute was part of a general attack on the idea that the different prize awarders should be bound by common rules. "We do not think," their counterproposal stated, "that the statutes should be drawn up in such a manner as to restrict the autonomy of the different prizeawarders at present or in the future, and to give one authority over the other."

The Karolinska delegates were for the most part successful in bringing home this point. In practice, it meant that the statutes came to contain only a few general rules concerning the prizes and the procedures for awarding them. Other matters were to be dealt with in the special regulations drawn up by each prize-awarding body and submitted to the government for final approval. This enabled the Karolinska delegates to delete the provisions in Lindhagen's draft statutes for a single Nobel institute organized in sections for physics, chemistry, and medicine. Instead, the final version of the statutes simply stated that each prize awarder would decide whether to establish an institute. Once created, the institute would be managed by the prize awarder in question while remaining the property of the Nobel Foundation. Other detailed provisions in Lindhagen's draft—for instance, those concerning the ranking of works according to their importance, the definitions of the fields in which the prizes were to be awarded, and the nominating system—met with a similar fate and were struck off the statutes.

By contrast, Lindhagen's formula for how to proceed if none of the works under examination possessed "the preeminent excellence . . . manifestly signified by the terms of the will" was included in the final version of the statutes. Here it was stipulated that a prize could be reserved for one year. If not awarded in the following year, the prize awarder could decide with a three-fourths majority to set it aside to form a special fund for the prize concerned. The proceeds of the funds could be used to promote the causes that Nobel had intended his prizes to further.

The promulgation of the statutes of the Nobel Foundation and the special regulations for distributing prizes from the Nobel Foundation by the government in 1900 closed the "organizing" phase in the history of the Nobel prizes in the Academy of Sciences. The statutes and regulations had created procedures and a machinery for awarding the prizes that even provided for the eventuality that prizes could *not* be awarded. However, these rules only provided the framework for the activity. The distinctive character of this activity would emerge from successive decisions.

Before discussing these decisions, a few words need to be said about the creation of the third award in the Academy: the prize for economic sciences established by the Bank of Sweden in memory of Alfred Nobel in 1968. When discussions started on the proposal to institute a prize in economics to commemorate the 300th anniversary of the Bank, the Academy was only one of several possible awarders, another being the Stockholm School of Economics (Handelshögskolan). During the discussions (1967–1968) between representatives of the Bank and those of the Nobel Foundation, in which leading Swedish economists also participated, it became clear that there were advantages in having the new prize awarded by the Academy of Sciences. First, the procedures for awarding the physics and chemistry prizes through nominations and recommendations by Nobel committees could easily be extended to the new prize. Second, the linkup with the science prizes would help alleviate the fears of those who were concerned about the epistemological status of economics as a "soft science" or "quasi-science." Third, the prominent Swedish economists, who were most willing to take on the task of awarding the prize, were all members of the Academy's economics, statistics, and social sciences section.

The situation in 1967 was very different, then, from that of 1897, as is shown by the dispatch with which the new prize was instituted. The secret negotiations, which had started in September 1967, were concluded in April 1968 when the Bank of Sweden made the formal offer to the Nobel Foundation to put a sum annually at the disposal of the Foundation that would be identical with that year's Nobel prize. Another sum corresponding to 65 percent of the prize money would be paid over to the Foundation

The Nobel Institute for Physical Chemistry, of which Svante Arrhenius became the Director. The building is still standing and is used by the students of Stockholm University.

to cover its own expenses and those of the Academy in connection with the award. By the end of May 1968, the offer had been accepted by the board of directors of the Nobel Foundation, the Academy of Sciences, the other prize-awarding institutions, and the Nobel family.

The new prize was kept separate from the existing ones in two principal ways: *one,* it was awarded "in memory of Alfred Nobel," and not given the status of an actual Nobel prize, and, *two,* it was given its own statutes and special regulations. This latter course was deemed preferable to changing either the statutes of the Nobel Foundation, which would hardly have been possible in any event since they were based on Nobel's will, or the special regulations for the award of Nobel prizes by the Royal Academy of Sciences. As with the original statutes and special regulations, those for the economics prize were submitted to the government for approval. They went into effect on January 1, 1969, and the first prize was awarded to Ragnar Frisch and Jan Tinbergen on December 10 the same year.[7]

Grossbetrieb in Science and the Nobel Institutes

At the turn of the century, scientists everywhere were aware of the growth in the scale of scientific research and teaching. In the 1880s and 1890s, German universities went on a building spree, which for most of them included the construction of new and larger institutes of physics and chemistry. In Sweden, the universities of both Uppsala and Lund were equipped with new facilities for physics and chemistry that were inaugurated during the first decade of the new century. Whatever were the causes of these changes—the ones most easily pinpointed were the increase in student enrollment and the emphasis on research both as part of training and as a necessary qualification for academic careers. They were surrounded by much rhetoric concerning the ineluctable march of the sciences towards ever greater organization and bureaucratization—the *Grossbetrieb* (grand enterprise) of science, as it became known. Even the great academies of sciences were affected; in 1899, when they joined together in the International Association of Academies, it was partly to launch collaborative projects, none of which came to fruition.[8]

The rhetoric undoubtedly allayed scientists' anxieties about new developments; in reality, although there was a change of scale, individual rather than collective research remained the norm until after World War I. In each major country, however, a few leading names in science seized the opportunity to forge alliances with powerful patrons willing to put up the money required to build and operate the facilities for *Grosswissenschaft.* The chain of institutes created under the aegis of the Kaiser-Wilhelm-Gesellschaft (1911), which was a product of the improbable alliance among the professoriate at Berlin University led by historian of religion Adolf Harnack, German industrialists, and the Kaiser, became the exemplar of these new facilities.[9]

The Nobel prizes were the antidote to these developments, for Nobel had conceived them as an aid to scholars, preferably younger ones, working alone in the heroic, nineteenth-century mold of a Pasteur or a Mendel. As discussed above, however, the representatives of the *scientific* institutions named in the will felt that the fortune of about 30 million kronor left by Nobel was so large that one-fifth of the interest on it seemed excessive for a prize, especially one functioning as a stipend. The endowment for the Nobel institutes, which at the outset amounted to 300,000 kronor for each institute, was thus wrested from the prize money. The uncertainty and contention that had characterized the protracted negotiations about the status and purposes of the institutes was not over, however, and it now delayed the process of setting them up. The problems involved were twofold:

First, an essential precondition of the *Grossbetrieb* of science was the pooling of resources and staff to set up good-sized research laboratories. The large Nobel institute envisioned by Arrhenius and Pettersson would have had institutional support from the Academy of Sciences, the Karolinska Institute, and Stockholm Högskola. The corporatist spirit in which the negotiations were conducted gave precedence, however, to the autonomy of each institution. The compromise, as we have seen, was to make all important decisions concerning the institutes, including whether they would be established, the prerogative of each institution. A further provision was inserted in the statutes, this time on the insistence of the executors, that prohibited *cumul,* that is, the staff of the Nobel institutes could not hold positions elsewhere without approval by the king-in-council. The prize awarders were also enjoined from using Institute funds for their own purposes. Added to the cautious and conservative attitude of the Academy mem-

Inaugural lunch when the first Nobel Institute was opened in 1909. Standing (left to right): David Gardner, "Technician" Barth, Johan Erik Johansson, Erich Marx, William Öholm, Harald Lundén, Victor Rothmund, Alexis Finkelstein, Leo Jolowicz. Seated: Carl Benedicks, Georg Bredig, Nils Ekholm, Svante Arrhenius, Wilhelm Ostwald, Max Reinganum, Karl Arrhenius.

bership, these different safeguards made for continued concern about how the operating costs of the Nobel institutes would be met.

Second, there was considerable confusion about the purposes of the institutes. Statutorily, they were justified by the scientific investigations that were to be carried out concerning the value of the discoveries proposed for the physics and chemistry prizes. In the minds of the scientists promoting them, however, the institutes were seen as carefree research establishments for prominent scientists. This view was espoused by Arrhenius when he wrote to Wilhelm Ostwald in 1898 that he himself would have a position at the Nobel institute and would be rid of his tiresome duties as president of the Högskola: "Perhaps that golden era will return when I only had to think about scientific matters."[10] Finally, there were the lingering hopes for the large Nobel institute envisaged by Arrhenius and Pettersson in 1898. A similar confusion concerning the purposes and expected results of independent research institutes prevailed in the memoranda leading to the establishment of the first Kaiser-Wilhelm Institutes a decade later.[11]

In the end, when the first Nobel institute was created, the overriding concern was to provide a research laboratory specifically for Arrhenius, Sweden's first Nobel laureate (Chemistry 1903), who might otherwise be lost to Swedish science. In 1904, Arrhenius announced that he had been invited by Friedrich Althoff, a powerful figure in the Prussian Ministry of Culture, to stand for election to the Berlin Academy of Sciences. Once he was elected, a chair would be created for him at the University of Berlin. The discussions that had dragged on for four years were brought to a rapid end by the Academy's decision to use part of the funds set aside for Nobel Institutes to create what became known as the Nobel Institute for Physical Chemistry, with Arrhenius as its director.[12]

Arrhenius's Institute was housed in temporary quarters until provided with a building of its own in Frescati north of Stockholm in 1909. The Academy of Sciences became its closest neighbor once the Academy's own new buildings had been completed in 1915. Built in the style of German research institutes, with one part housing the laboratory and the library and another providing comfortable living-quarters for the director and his family,[13] the Arrhenius Institute was a far cry from the "big" institute envisaged in proximity to the Högskola on Observatory Hill. Still, given the poverty of research facilities in Stockholm, the Institute with its two staff positions (the director and his assistant), its laboratory space for visitors, and its own series of bulletins (*Meddelanden från Kungl. Vetenskapsakademiens Nobelinstitut,* Volumes 1–6, 1905–1929) was a welcome addition.

The first decade of the Institute's functioning cleared up some of the confusion that had surrounded the idea of Nobel institutes. First, it became manifest that the Nobel committees could dispense with the scientific investigations of the works proposed, since those seriously considered for the prizes had already been experimentally verified. In the period up to World War I, Arrhenius was called on only three times to examine proposed works, all of them involving the candidacy of Walther Nernst.[14] Second, while Arrhenius remained productive until his death in 1927, his later work in cosmic physics and immunotherapy, although highly original, did not have the same influence as his theory of electrolytic dissociation (1887). Hence, it could not be used to prove the proposition that a prominent scientist would produce important results if given the time and resources to devote himself entirely to research. Here, again, the comparison with the Kaiser-Wilhelm Institutes is instructive since the nobelized or nobelizable scientists who headed these often brought to their institutes prestige rather than new scientific discoveries.[15]

The question of establishing new Nobel institutes received only intermittent attention until 1927, when Arrhenius's death confronted the Academy with the problem of apportioning his institute.[16] The committee of four, which became known as the Nobel Institute Committee, that the Academy hastily appointed to study the matter took its time.[17] After much consultation with the Nobel Committees for Physics and Chemistry, and with the pertinent sections of the Academy (the third and fourth), it finally presented the Academy with a set of proposals in 1932. These were threefold: 1) that the Academy establish a Nobel institute for chemistry; 2) that the buildings of Arrhenius's institute be used for this purpose; and 3) that the Academy establish a Nobel institute for theoretical physics, but that no new building be erected for this purpose.[18] Of the three recommendations, the Academy only adopted the third and then not until 1933.[19] The chief reasons for this course of events—in general, the issues were many and complex especially as they involved competition for resources between and within the Nobel Committees for Physics and

Chemistry—were the problems of how the institutes would be financed and who would head them.

The cost of operating the Nobel Institute for Physical Chemistry had been a major worry of Arrhenius's in the last years of his life when postwar inflation ate into the annual allocation of funds—around 40,000 kronor in 1923—from the Academy. The decline in the proceeds of the different Nobel funds is best illustrated by the prize money itself, which in 1923 had fallen to 115,000 kronor, the lowest amount ever.[20] By the end of the 1920s, however, the financial prospects had brightened, particularly for the chemists, who seriously envisaged taking over the Arrhenius Institute and maintaining it as the Nobel institute for chemistry.

The grounds for optimism lay in the significant increase in the chemists' special fund, which had benefitted from four reserved prizes since 1916. In 1930, the capital of around 600,000 kronor yielded an annual income of 28,000 kronor. Although part of the latter sum could be allocated to maintain the planned Nobel institute for chemistry—comparable use was already being made of the funds of the literature and peace prizes—there was a serious obstacle to this proposal. Starting in the early 1920s, the Nobel Committee for Chemistry—and as we will see shortly, that for physics as well—had fallen into the habit of using about two-thirds of the income annually as grants for the research projects of Swedish chemists. This use of the proceeds of the special funds was in accordance with paragraph 5 of the statutes in that the grants could be held "to promote the objects which the testator ultimately had in view in making his bequest, in other ways than by means of prizes." Although the sums involved were not big, the largest grants amounting to about 10,000 kronor, they were more easily available and freer of "red tape" than the regular funds of university laboratories and could thus be used to purchase research equipment. Although internationally well-known scientists, such as The(odor) Svedberg and Hans von Euler benefitted the most, younger ones, such as Hans Pettersson, were also among the recipients.[21]

As the different documents appended to the report of the Nobel Institute Committee show, the chemists disagreed about whether it was preferable to maintain the grant program or to allocate these resources a Nobel institute for chemistry. In a dissenting opinion to the committee report, Lars Ramberg, professor of chemistry at Uppsala University, argued that the income from the special fund would be used more efficiently as grants benefitting already existing laboratories than if it were allocated to a Nobel institute for chemistry, which in any event would not have up-to-date facilities. The opposite opinion was advanced by Wilhelm Palmaer, professor at the Royal Institute of Technology and former secretary of the Nobel Committee for Chemistry.[22] In the end, the Academy followed Ramberg's opinion and decided not to establish a Nobel institute for chemistry.[23] The problem of who would head the institute no doubt contributed to this decision, for although Palmaer coveted the post, prominent Swedish chemists such as von Euler and Svedberg do not seem to have been tempted.

In contrast with chemistry, the circumstances in physics with respect to both financing and leadership favored the establishment of a Nobel institute. This became the Nobel Institute for Theoretical Physics, created by decision of the Academy in 1933 with Carl Wilhelm Oseen as its director and located in the old Arrhenius Institute.[24] The creation of the institute did not hinge on the physicists' having more funds on hand than the chemists— quite the contrary, since by 1933 only two Nobel prizes had been allocated to the special funds as against four in chemistry. The physics committee also gave grants to university physicists but on a much smaller scale; on the average, the annual amount in the 1920s was about one-third of that paid out in chemistry. In proposing the establishment of an institute, Manne Siegbahn and Henning Pleijel, both members of the Nobel Committee for Physics, could argue convincingly, however, that the main cost of the institute would be the director's salary. If any experimental work were to be carried out, this could be financed from the special fund on a project basis.[25]

The activities of the Nobel Institute for Theoretical Physics were those of its director, C.W. Oseen, who remained in this position, helped by an assistant, until his death in 1944. Oseen was trained in mathematics and mathematical physics at Lund University where he taught until 1909, when he was appointed professor of mechanics and mathematical physics at Uppsala. By then, he had sojourned twice at continental universities: at Göttingen in the winter of 1900–1901, where he was much taken by David Hilbert's work on partial differential equations, and at Paris in 1907. His travels and extensive reading also introduced him to the latest advances in theoretical physics, an interest that he pursued alongside the work in hydrodynamics, particularly anisotropic liquids, that represented his main contribution to mathematical physics. At Uppsala, Oseen gave his first set of lectures on the theory of heat radiation, organized a graduate seminar on atomic theory where Niels Bohr's work was at the center of attention, and generally stimulated his students to do advanced work in theoretical physics. During his time at Uppsala, he directed eight theses in this area, thereby training most of the first generation of Swedish theoretical physicists.[26]

Oseen had put his extensive knowledge of theoretical physics at the service of the Nobel Committee for Physics before he became a member of the committee in 1923. As a special expert to the committee in 1922, he wrote the reports on Einstein's law of the photoelectric effect and Bohr's atomic theories. These opened the door for the Academy's decision of that year to award Einstein the prize reserved in 1921 "for his services to theoretical physics, and especially for his discovery of the law of the photoelectric effect," and Bohr the prize of 1922 for his atomic theories. Oseen's intervention stemmed the swelling tide of nominations citing Einstein for his relativity theory, which the experimentalists, who constituted the majority of the committee, found difficult to accept since they felt it lacked sound proof.[27]

Throughout the 1920s, Oseen was for all practical purposes the only representative of theoretical physics on the committee, and hence the one who followed developments in atomic theory and quantum mechanics most closely. He wrote the reports on, among others, Werner Heisenberg, Louis and Maurice de Broglie, James Franck, Gustav Hertz, Erwin Schrödinger, and Arnold Sommerfeld.[28] It seemed right and proper then that when he began to feel burdened by his teaching duties at Uppsala, an institute would be created for him where he could devote himself wholly to research and writing. This meant that for its second Nobel Institute the Academy had opted for the same "personalized" formula as for the first. In giving the directorship to Oseen, the idea that the Institute should provide assistance in the investigations necessary to make the award was also being realized, albeit in a different manner from that envisaged when the statutes were drawn up.

In 1936 the conditions were finally at hand for establishing the large Nobel institute for experimental research that had eluded those, such as Arrhenius and Pettersson, who pioneered the idea. The scientific entrepreneurship and vision necessary to organize and finance a large laboratory were no doubt the most important of these conditions. They were provided by Manne Siegbahn, professor of experimental physics at Uppsala University and winner of the 1924 Nobel prize for physics for his precision work in X-ray spectroscopy.[29] By the late 1920s, Siegbahn had come to feel that he could not realize his ambition for an institute with full-time researchers and a technical staff skilled in constructing instruments—in short, the *Grossbetrieb* of science—within the university setting. Initial plans laid in the early 1930s, according to which the institute would be placed under the Academy but be operated with private funds secured mostly from the Wallenberg and Rockefeller Foundations, were cut short by the Great Depression. Siegbahn persisted, and the scheme that he contrived in 1936 was accepted by the different partners in the enterprise. They were three: the Academy of Sciences and its Nobel Committee for Physics, which provided the money (around 800,000 kronor) for construction from the funds set aside in 1901 for the Nobel institute for physics; the Knut and Alice Wallenberg Foundation, which contracted with the Academy to meet the bulk of the operating costs (55,000 kronor out of an estimated 90,000 kronor); and the Swedish government, which instituted a personal professorship for Siegbahn at the Academy and undertook to meet the remainder of the operating costs.[30]

The Academy of Sciences Institute for Experimental Physics, which opened in 1937, represented in many ways the fulfillment of the original ambitions for the Nobel institutes; hence, it was proper that it became known as *the* Nobel Institute. Although the institute would not have got off the ground without the support of the Wallenberg Foundation, the government's creation of a personal professorship for Siegbahn was perhaps more important. This was not just because it was the *sine qua non* for the functioning of the institute, but because it represented *public* support for the idea of providing prominent scientists with the means to devote themselves wholly to research. The arguments of Social Democratic minister of education, Arthur Engberg, who supported Siegbahn in the face of strong opposition from university corporations, that of Uppsala in particular, are oddly reminiscent of those advanced in favor of the Kaiser-Wilhelm Institutes earlier in the century. In this sense, the creation of the Nobel Institute closed a chapter in the history of scientific institutions. It also presaged the growth of government funding of scientific research in the postwar period through a panoply of mechanisms, among others the Atomic Energy Committee (1945) and the Natural Science Research Council (1946). The Nobel Institute would ultimately be merged into this system when the government took over responsibility for its financing in 1964.[31]

The Significance of the Nobel Prizes for the Academy and Swedish Science

In a small scientific community, such as Sweden's, most scientists with a national reputation will sooner or later hold membership in the Academy of Sciences. This makes it difficult to separate the effects of the prizes on the Academy as such from those that pertain directly or indirectly to the disciplines of physics and chemistry. However, an attempt to do so will be made in the following paragraphs.

The effects of the prizes on the Academy as such can be observed most directly in the organization into sections *(klasser)*. In the long history of the Academy, the sections for natural history, medicine, practical arts, and economic science had overshadowed those for physics and chemistry (combined with mineralogy). The need to strengthen the two sections with primary responsibility for Nobel prize selections made the Academy decide in 1904 to increase the physics (combined with meteorology) section from six to ten members and to separate the chemistry section, also with ten members, from the mineralogy one. Through subsequent changes in the statutes of the Academy, the membership has been increased numerically from 100 at the turn of the century to nearly 300 at present. This has made it possible to ensure that the members of the "Nobel sections" of physics, chemistry, and economics, statistics, and social sciences represent the broad range of subject areas necessary for Nobel prize selections.

The description in the main part of this article should leave no doubt that Nobel's legacy was a windfall for physics and chemistry in Sweden. It permitted leaders in these disciplines—Arrhenius, Oseen, and Siegbahn, among others—to realize their ambitions to set up research institutes that were independent of the universities. For Arrhenius, who was something of a renegade, his institute permitted him to work in areas—cosmic physics and cosmology, for instance—for which there was no obvious niche in the universities. The grants from the special funds of the Nobel Committees for Physics and Chemistry benefitted physicists and chemists, who were not directly involved with the Nobel Institutes. The sums involved were not very large; in physics, the total amount from 1919 to 1939 was 147,020 kronor, and in chemistry, 329,117 kronor, that is, sums approximately corresponding to one and two Nobel prizes respectively. These grants came in addition to regular departmental budgets, however, and could be obtained rapidly without too much paperwork. In this respect, the grant program run with the Nobel funds was akin to that of the *Notgemeinschaft der deutschen Wissenschaft* that helped tide German science over the difficult years of post-World War I inflation.[32]

The funds put at the disposal of the Academy through the Nobel legacy provided Swedish scientists with experience in science policy decision-making long before this term had gained currency. The needs for Nobel institutes were justified in lengthy arguments that took into account the state of research and teaching in the fields concerned, the advisability of freeing university professors from their teaching duties, and the functioning of independent research institutes abroad. To give some examples:

—The proposal by Oseen and Pleijel for a Nobel institute for theoretical physics without research facilities was accompanied by an extempore survey of changes in the laboratory equipment and needs of German theoretical physicists from the 1880s to the 1930s. Recalling the large facilities erected for the mathematical physicists Hermann von Helmholtz and Woldemar Voigt in the former period, they went on to describe the present situation; in Göttingen, Max Born only used a fraction of Voigt's large institute, and in Munich, Arnold Sommerfeld saw no use for his large laboratory facilities. By contrast, it was important to provide rooms for conferences of the kind held regularly at Bohr's institute in Copenhagen or Debye's in Leipzig.[33]

—In his dissenting opinion to the Nobel Institute Committee recommendation for a Nobel institute for chemistry, Ramberg argued that chemistry was represented by 20 chairs at Swedish universities—double the number existing in physics—and that the discipline, which already had difficulty in regenerating itself, would be not be well served by draining off personnel to a Nobel Institute.[34]

—Finally, Siegbahn's extended battle for his own institute involved him in lengthy arguments about the cost-effectiveness of independent research institutes compared to university teaching laboratories.

In these and similar ways, Swedish scientists and the Academy of Sciences gained experience that prepared them for the advent of "big science" after World War II and the science policy decisions that this would entail.

Looking further afield, there is no doubt that one of the main roles of the Nobel prizes in Swedish science was to hasten its internationalization. The prizes burst upon a small and peripheral scientific community dominated by national concerns: on the one hand, the Linnean tradition of natural history, and, on the other, the demands that the industrialization and modernization of the country put on its scientists.[35] By increasing the weight of basic research in physics and chemistry in the national scientific enterprise, the prizes went counter to both these tendencies. More important, they helped bring disciplinary developments in Sweden in step with those of the major science-producing countries, Germany in particular. The nomination process and the annual Nobel ceremonies expanded the international horizons of Swedish scientists, who were given opportunities to visit laureates' and— even more obviously—candidates' laboratories. These were eagerly seized on by the vanguard of internationally minded scientists—Arrhenius, Svedberg, Oseen, and Siegbahn—whose students and collaborators then followed suit.

Notes

1. Elisabeth Crawford, *The Beginnings of the Nobel Institution: The Science Prizes, 1901–1915* (Cambridge, 1984), Ch. 1, 10–30.
2. Ragnar Sohlman, *The Legacy of Alfred Nobel: The Story behind the Nobel Prizes* (London, 1983).
3. In terms of its purchasing power at the time, the Nobel prize of 1901 (150,000 kronor) was approximately double that of 1988 (2.5 million kronor).
4. Letter from A. Nobel to A.E. Nordenskiöld, March 15, 1984, Nobel Collection, Swedish National Archives. Nobel's foreign Academy membership could have been used by the Swedish relatives in their efforts to prove that Nobel was domiciled not in Sweden as the execu-

tors and their attorney claimed, eventually with success, but in France. They had good reason to believe that a French court would declare the will invalid since it did not meet the strict requirements of the Code Civil, which stipulated that only formally constituted institutions could receive bequests.

5. The account that follows is based on Crawford, *The Beginnings*, Ch. 3, 60–86. The statutes of the Nobel Foundation (1900) and the special regulations concerning the distribution of prizes from the Nobel Foundation by the Royal Swedish Academy of Sciences are reproduced as Appendix B.

6. For examples of how the proceeds of prize funds were used for the general benefit of the French Academy of Sciences, see Maurice Crosland, "From Prizes to Grants in the Support of Scientific Research in France in the Nineteenth Century: The Montyon Legacy," *Minerva* 17:3, Autumn 1979, 355–380.

7. Sveriges Riksbank, "Donationsbrev," June 6, 1968; *Stadga för Sveriges Riksbanks pris i ekonomisk vetenskap till Alfred Nobels minne* (1969); *Stadga med särskilda bestämmelser angående utdelning genom Kungl. Vetenskapsakademien av Sveriges Riksbanks pris i ekonomisk vetenskap till Alfred Nobels minne* (1969).

8. David Cahan, "The Institutional Revolution in German Physics, 1865–1914," *Historical Studies in the Physical Sciences* 15:2 (1985), 1–65; Brigitte Schroeder-Gudehus, "Division of Labor and the Common Good: The International Association of Academies, 1899–1914," in *Science, Technology and Society in the Time of Alfred Nobel*, eds. C.G. Bernhard, E. Crawford, and P. Sörbom (Oxford, 1982), 3–20.

9. Elisabeth Crawford and John L. Heilbron, "Die Kaiser-Wilhelm-Institute für Grundlagenforschung und die Nobel-Institution," in *Forschung im Spannungsfeld von Politik und Gesellschaft: Zum 75jährigen Bestehen der Kaiser-Wilhelm-Max-Planck-Gesellschaft (1911–1986)*, eds. R. Vierhaus and B. vom Brocke (Stuttgart, 1989) (in press).

10. Hans-Günther Körber, ed., *Aus dem wissenschaftlichen Briefwechsel Wilhelm Ostwalds*, Vol. II (Berlin, 1969), 151.

11. Crawford and Heilbron, "Die Kaiser-Wilhelm-Institute."

12. Nobelprotokoll, KVA, 3dje och 4de klasser, November 12, 1904; Nobelprotokoll, KVA, November 26, 1904; Hans von Euler, "Svante Arrhenius, 1859–1927," in *Swedish Men of Science, 1650–1950*, ed. S. Lindroth (Stockholm, 1952), 226–238.

13. At present the old Arrhenius Institute building houses the Student Union of Stockholm University, and the director's home is a student cafeteria.

14. Crawford, *The Beginnings*, 154–155.

15. Crawford and Heilbron, "Die Kaiser-Wilhelm-Institute."
16. Robert M. Friedman, "Americans as Candidates for the Nobel Prize: The Swedish Perspective," in *The Michelson Era in American Science 1870–1930*, eds. S. Goldberg R. Steuwer (New York, 1988).
17. The members of the Nobel Institute Committee were: E. Hulthén and M. Siegbahn representing the Nobel Committee for Physics, and H. von Euler and L. Ramberg the Chemistry Committee.
18. Nobelprotokoll, KVA, November 23, 1932 (Bil. Kommittéutlåtande: Till Vetenskapsakademien, October 8, 1932).
19. Nobelprotokoll, KVA, May 10, 1933.
20. *Nobel Foundation Directory, 1987–1988* (Stockholm, 1987), 15.
21. Nobelprotokoll, KVA, 1920–1938.
22. Nobelprotokoll, KVA, November 23, 1932 (Bil. L. Ramberg, Särskild skrivelse, October 8, 1932; W. Palmaer, Yttrande angående inrättandet av Nobelinstitutets avdelning för kemi, May 2, 1930, and Till Nobelinstitutions kommittén, November 2, 1931).
23. Nobelprotokoll, KVA, May 10, 1933. From 1933 until World War II, a chemistry institute of sorts was functioning at the Arrhenius Institute Building in that it housed the laboratory where W. Palmaer and his assistants were engaged in research on the corrosion of metals.
24. Nobelprotokoll, KVA, May 10, 1933.
25. Nobelprotokoll, KVA, November 23, 1932 (Bil. M. Siegbahn and H. Pleijel, Till Vetenskapsakademiens Fysiska Nobelkommitté, January 31, 1930; C. W. Oseen and H. Pleijel, Till K. Vetenskapsakademiens Nobelinstitutskommitté, May 9, 1932).
26. Ivar Waller, "Carl Wilhelm Oseen," *Levnadsteckningar över Kungl. Svenska Vetenskapsakademiens ledamöter*, Vol. 8 (1949–1954), 121–143; Lamek Hulthén, "C.W. Oseen: Minnestal i Göteborgs Kungl. Vetenskaps- och Vitterhetssamhälle, den 24 januari 1945," in *Bihang till Göteborgs Kungl. Vetenskaps- och Vitterhetssamhälles Handlingar* 64 (1945), 1–4.
27. Elisabeth Crawford and Robert M. Friedman, "The Prizes in Physics and Chemistry in the Context of Swedish Science: A Working Paper," in *Science, Technology and Society in the Time of Alfred Nobel*, 311–331.
28. V. Carlheim-Gyllensköld, who served on the committee from 1910 until his death in 1934, was the only other member with a knowledge of mathematical physics. For excerpts from some of Oseen's reports, see Ingmar Bergström, "Manne Siegbahn and the 1924 Nobel Prize for Physics," *Physica scripta* T22 (1988), 9–20.
29. Ibid.
30. Olle Edqvist, "Manne Siegbahn," *Kosmos* 64 (1987), 163–173; Robert M. Friedman, "Manne Siegbahn," *Dictionary of Scientific Biography*

(forthcoming); "Statsutskottets utlåtande Nr. 64," *Bihang till riksdagens protokoll 1936,* 6 saml., 27 pp.; Nobelprotkoll, KVA, April 22, 1936.

31. Edqvist, "Manne Siegbahn," 175–176.
32. John L. Heilbron, *The Dilemmas of an Upright Man: Max Planck as Spokesman for German Science* (Berkeley, 1986), 90–93; Friedrich Schmidt-Ott, *Erlebtes und Erstrebtes, 1860–1950* (Wiesbaden, 1952). For the sums in physics and chemistry, see Per Santesson & Wilhelm Pehrsson, *Översyn av Nobelstiftelsens grundstadgar: Utlåtande avgivet den 30 september 1968,* Nobelstiftelsen (Stockholm, 1968).
33. Nobelprotokoll, KVA, November 23, 1932 (Bil. M. Siegbahn and H. Pleijel); Christa Jungnickel and Russell McCormmach, *Intellectual Mastery of Nature: Theoretical Physics from Ohm to Einstein,* 2 vols. (Chicago, 1986).
34. Nobelprotokoll, KVA, November 23, 1932 (Bill. L. Ramberg).
35. Crawford, *The Beginnings of the Nobel Institution,* 30–42; Gunnar Eriksson, *Kartläggarna: Naturvetenskapenstillväxt och tillämpningar i det industriella genombrottets Sverige, 1870–1914,* Acta Universitatis Umensis, No. 15 (Umeå, 1978). (Summary: The Growth and Application of Science in Sweden in the Early Industrial Era, 1870–1914).

CARL GUSTAF BERNHARD

Research Institutes

UNLIKE many other academies, the Academy of Sciences sponsors research institutes, although unlike the academies of Eastern Europe, which are responsible for the major research activities in their countries, it runs only a limited number. This has been so ever since the early life of the Academy in the eighteenth century. As a rule the first steps towards the formation of these offshoots were taken in response to current needs as they arose. In many cases the institutes concerned grew fast and expanded their activities, thus fitting better into a university organization or growing into an organization of their own. This meant that the Academy was repeatedly left free to start new projects without the heavy burden of administering too many large institutes.

Earlier Institutes

Only a few weeks after the founding of the Academy in 1739, Carl Linnaeus presented his major work *Hortus Cliffortianus,* which became the start of the Academy library. During the eighteenth century, the library received many volumes as donations, and in 1749 exchanges began with other academies, the Royal Society in London being the first partner. The secretary of the Academy was also its librarian until the 1820s, when Berzelius saw to it that the library was given its own appropriation and its own staff. During the late nineteenth century, it grew into the most important in Scandinavia. With purchases, donations, and exchanges, the library expanded until in 1978, during the librarianship of Wilhelm Odelberg, its total stock was estimated at 15,000 shelf meters. It possesses notable special collections, including first editions, manuscripts, and volumes of prints dating back to before the Academy was founded. On July 1, 1978, by agreement with the government, the Library of the Academy was transferred to Stockholm University, where it came to form a mathematics and science section, while still retaining its own name.

In 1753 the Academy's Astronomical Observatory was opened, with Pehr Wargentin as its first director. It was not replaced until a new observatory was built at Saltsjöbaden in 1931; this was transferred to Stockholm University in 1973 (see Ulf Sinnerstad's article in this volume).

The Academy's Institute of Physics, which was founded in 1849, remained active until the death of its last director, the Academy physicist Bernhard Hasselberg, in 1922. Its valuable collection of historical instruments formed the nucleus of the Academy's Museum of the Exact Sciences.

The young Academy soon began to receive contributions to a natural history collection, including a large wasp's nest from Queen Lovisa Ulrika. In 1819 Marshal of the Court Gustaf von Paykull donated a large zoological collection to the state, and this formed the basis of the Swedish Museum of Natural History. During Berzelius' term the collections were steadily enlarged, partly as a result of voyages of exploration to different parts of the world. As the collections grew, so did the amount of research work done. The need for new premises became increasingly acute, and in 1916 the Museum moved into its present building in Frescati. The Museum had by now acquired an ethnography section. This was hived off from the other sections in 1935 to become the Swedish Ethnographical Museum, still under the supervision and care of the Academy. Finally, in 1964, both the Museum of Natural History and the Ethnographical Museum were brought under governmental control (see Gunnar Broberg's article in this volume).

Another of the Academy's important innovations was the commencement in 1859 of meteorological observations at stations in different parts of Sweden. These activities led to increased international co-operation in meteorology, and in 1873 the Central Meteorological Office of Sweden was set up under the Academy's superintendence. Following rapid expansion and amalgamation with the National Hydrographic Office, the State Department for Meteorology and Hydrology was opened in 1919, later becoming the Swedish Meteorological and Hydrological Institute (SMHI).

Alfred Nobel's large donation made it possible to use Nobel funds to open research laboratories, which became known as Nobel institutes. In 1905 the Nobel Institute for Physical Chemistry was founded under the direction of Svante Arrhenius, who remained in charge of it until his death in 1927. It then closed down, to be replaced by the Nobel Institute for Theoretical Physics, established in 1933 with C.W. Oseen as its director. In 1936 the Riksdag granted money for a chair in experimental physics, intended for Manne Siegbahn. A laboratory building was also erected with the aid of a Nobel grant, and with this the Research Institute for Experimental Physics was born. After many years of successful activity, this institute was reorganized in 1964 and became the state-controlled Research Institute for Physics. Meanwhile, the Nobel Institute for Theoretical Physics had also been reorganized, first as the Laboratory of Organic Chemistry under the directorship of Bror Holmberg (1943) and, after his retirement, as the Nobel Institute, Department of Chemistry (1951). There are currently two Nobel institutes, for physics and chemistry, as organizations with liaison and representative duties: sending invitations to guest researchers and speakers, arranging symposia, conferences, etc. (see Elisabeth Crawford's article in this volume).

The Kiruna Geophysical Observatory was established by the Academy in 1957 after ten years of planning. This was a development of the work in geophysics that had been carried on at the Academy's Natural Science Station in Abisko since 1912. Activities were transferred to provisional premises in Kiruna in 1948, for ionospheric soundings under the leadership of Olof Rydbeck and to register terrestrial magnetism under the supervision of Nils Amboldt. With support from the government and the local municipality, a permanent observatory building was built near Kiruna and opened in 1957. In the same year, Bengt Hultqvist became director of the observatory, and the subsequent collaboration with Umeå University was marked in 1967 by the appointment of Hultqvist as professor of geocosmophysics there.

Under Hultqvist's dynamic guidance, research into geocosmophysics developed rapidly at the observatory. The main field of interest was those phenomena that are characteristic of the upper atmosphere in the auroral zones and the related processes in the magnetosphere. On July 1, 1973, the Kiruna Geophysical Observatory acquired the status of an autonomous research institute directly under the Office of the Chancellor of the Universities and

Colleges of Sweden. The Kiruna Geophysical Institute (as it is now known) is also responsible for research education at Umeå University (see below on the Abisko Natural Science Station).

The Bergius Foundation

The Bergius Foundation came into being in 1791, when the Academy received the bequest of Peter Jonas Bergius, which consisted of the property of Bergielund, situated northwest of the city of Stockholm. He had owned it jointly with his brother Bengt, who had died in 1784; inspired by a profound interest in natural history and horticulture, they had turned the property into a botanical institute. This meant that the donation also included one of the largest private libraries of the period, a herbarium of 15,000 sheets, and a well-tended garden, in which medicinal plants, vegetables, fruit, berries, and tobacco were cultivated. The aim of the Foundation was, in the words of the will, to establish and maintain a school of horticulture, run a market garden, and carry out botanical research. The Foundation thus became the property of the Academy, with its own board and a botanist of professorial status as director.

In 1791 Olof Swartz became the first Bergius professor and subsequently also the permanent secretary of the Academy, which he remained until his death in 1818. He had previously devoted himself to the flora of the West Indies, whereas his successor, Johan Emanuel Wikström, remained nearer to home and is best remembered for his work on the flora of Stockholm. In 1856 Nils Johan Andersson became head of the Foundation after taking part in the round-the-world voyage of the Swedish frigate *Eugenie* (1851–1853) and publishing an account of the vegetation of the Galapagos Islands, virtually unknown at that time, which contained a description of new species.

With the growth of Stockholm, Bergielund was sold, and in 1885 activities were moved to an area purchased in Frescati, where the Bergius Garden and the Bergius Institute are still situated. It was under the vigorous directorship of Veit Wittrock (1879–1914) that buildings were erected, gardens laid, and the school of horticulture enlarged, while the present institute building and the greenhouse for tropical terrestrial plants were built under his successor, Robert Fries (1915–1944).

For both economic and practical reasons, the school of horticulture was eventually closed, and in 1969 the botanical garden was sold to the state to be transferred to Stockholm University. In 1979 the commercial section was put out on lease. The Foundation retained—except for the commercial section—the Bergius Institute built in 1936. The director of the Institute is also the director of the botanical garden, which means that collaboration between the Foundation and the Bergius Garden continues, although the latter now belongs to the university. This collaboration has chiefly concerned research into plant ecology with reference to man's influence on the land and on horticulture.

As Bergius professor, Veit Wittrock specialized in multiform plants, and his studies were combined with attempts at cultivation in the Garden. His best-known works are those on various *Viola* species and on *Linnea borealis*. His successor, Robert Fries, who took part in a number of expeditions to South America and Africa, concentrated both on plant geography and morphology and on taxonomy. His most widely known publications are in phanerogam taxonomy, and after his expedition to the mountains of East Africa, he described over 300 new species and proposed seven new genera. The main field of research of his successor Rudolf Florin was the morphology and taxonomy of the gymnosperms, including our common conifers, and he made particular use of fossil material.

The earlier work of Måns Ryberg, director from 1970 to 1983, was mainly in morphology and taxonomy, particularly with relation to the *Fumariaceae*. Under his leadership research efforts were focussed on semi-natural land produced by changes in agriculture and on autoecological studies of certain plant genera. Taxonomic research was carried out in certain groups of higher and lower plants. In horticulture Ryberg took an active part in the work of the Swedish section of the Nordic Gene Bank on vegetatively propagated garden plants, particularly strawberries.

In recent years Måns Ryberg has been working on the composition and developmental history of the vegetation of deciduous forests, with his interest turning increasingly to ecology, plant geography, and nature conservancy. The research directed by Ryberg at the university has dealt with the life history of different plants and how this has been affected by various abiotic and biotic factors.

Some realignment of the direction of research from ecology to taxonomy took place in 1983, when the domain of the professorship was defined as "taxonomy, plant geography and cytogenetics." Bengt Jonsell, who succeeded Måns Ryberg as director, had previously done much work on the biosystematic problems of the flora of Scandinavia and on tropical plant taxonomy, particularly concerning the *Cruciferae*. He completed a taxonomic phytogeographical monograph on the genus *Farsetia,* for which the Horn of Africa is a major center of species formation.

One of the Institute's recent projects is in biosystematics; it deals with the dynamic evolutionary characteristics of the Baltic land uplift and also includes studies of the conditions and status of endangered plant species.

Since the 1970s the assistant professor Lars Erik Kers has been carrying out floristic field studies of different ecosystems in Sweden with special attention to various fungi. In 1983 he published his systematic-floristic works on truffles and on the ethylene production of their sporophores. He also undertook a revision of the *Capparidaceae* family for tropical floras, particularly that of Gabon. With the contributions of Kers and Jonsell on *Capparidaceae* and *Cruciferae* respectively, the Institute played a significant role in the production of *Flore du Gabon* and *Flora malesiana,* which were completed in 1987, and of the flora of Ethiopia, which is still in production.

When the Bergius Garden celebrated its centenary in August 1986, a Nordic symposium on "Biosystematics of the Nordic Flora," which covered the history of the immigration of various plants, was held at the Academy of Sciences under the chairmanship of Jonsell. The symposium played an important role in the start in 1987 of work on a new scientific vascular flora of the five Nordic countries, *Flora Nordica.*

The plant world of the Nordic countries is changing rapidly, and these changes will be illustrated in the new flora. There are also matters of taxonomic interest, such as variation within species, hybridization, reproduction, and evolutionary and ecological problems. The *Flora* is to appear in four volumes and to be ready by the end of the 1990s. The chief editorial office is at the Bergius Institute.

Kristineberg Marine Biological Station

In August 1835 the secretary of the Academy of Sciences, Jacob Berzelius, received a letter from the curator of the Museum of Natural History, Bengt Fries, who wrote: "It was most trying not to be able to tarry longer by this rich fount of creatures, where the water teems with living things, each one more strange than the last." He had been at Kristineberg on the island of Skaftö, at the mouth of Gullmar Fjord, and the subject of his enthusiasm was the marine fauna of the fjord. This threshold fjord, some 15 miles long, which is about 35 meters deep in the middle part, has an unusually rich fauna and flora. Among the 1,500 deep-basin species described, many have to be sought at a depth of 300 meters in the Skagerrak.

While Fries was the person who drew attention to the promise of Gullmar Fjord for marine biology, it was Sven Lovén who created a center there for zoological, botanical, physiological, and anatomical research. After completing a scientific expedition to Spitzbergen in 1837, he had devoted himself to the animal life of the Bohuslän coast, and in 1839 he came to Kristineberg, where he studied a series of lower marine animal forms. This attracted many zoologists and botanists to Kristineberg, and the place earned international renown.

Lovén's wish to set up a permanent research station was realized when in 1876 the Academy received a donation from Dr. Anders Fredrik Regnell "for the establishment of a zoological station on the west coast of Sweden." The first property was purchased in 1877, and serious work started the following year. Activities expanded rapidly, and the need for more space was soon felt. Laboratories, pumping house, water reservoirs, and more living accommodation were gradually purchased or built, thanks largely to generous donations by Emil and Anna Broms, which also made it possible to build the first research vessel, an oak double-ender named "Sven Lovén" that served the station for 70 years. Gradually other vessels were acquired, both for studies in the fjord itself and to collect material for laboratory work.

For the centenary of the station in 1977, the old boats were replaced by two newly built, modern research vessels for sea and fjord work. With the help of donations from Erna and Victor Hasselblad and a grant from the Wallenberg Foundation, the 26-meter *Arne Tiselius* was built for deep-sea work, while a grant from the Mary von Sydow (née Wijk) Donation Fund enabled a 12-meter aluminium vessel, the *Oscar von Sydow,* to be acquired for work in the fjord. The latest addition to the fleet is a robot, the *Christine,* for underwater work.

The station has always been open to foreign guest researchers, who, like Swedish scientists, make use of the particular conditions offered by the local environment. One of the beneficial results of this activity has been the charting of the hydrographic, zoological, and botanical characteristics of Gullmar Fjord, further enhancing the value of the area for marine research. Both Swedish and international organizations have recommended the area as a nature reserve on the grounds of its distinctive character and its great scientific importance.

Sven Lovén served as director until 1882, when he was succeeded by Hjalmar Théel, who in 1907 published a list of all the marine animal species found at Kristineberg. After the building of the winter laboratory, which enabled work to continue throughout the winter, a post was created in 1906 for a resident director, and its first holder was Hjalmar Östergren. During the 1920s the director was Magnus Aurivillius, after which Gunnar Gustafsson gave long and highly valued service from 1931 to 1959 as director and also as leader of university courses in marine biology. His successor, Bertil Swedmark (1959–1975), was one of the pioneers in meiobenthology and attracted foreign researchers and students in this field. He was succeeded in 1975 by the present director, Jarl-Ove Strömberg.

Most of the research work at the station takes place in the Gullmar Fjord, but it also extends to the Skagerrak, the Kattegat, and the Sound. The waters around Kristineberg are functionally a part of the marine ecosystem of the whole of this area, which is dependent on hydrographic conditions in the transitional zone between the Baltic and North Seas. The brackish water of the Baltic flows via the Sound and the Belts out into the Kattegat, is forced by the rotation of the earth against the west coast of Sweden, and follows this up into the Skagerrak. Due to mixture with the underlying more saline water of the North Sea, the average salt content gradually rises in the surface current which represents the outflow from the whole Baltic Sea and from the urban areas of Sweden's west coast and the Danish islands.

The hydrographic conditions of Gullmar Fjord itself were described around the turn of the century in the work of the pioneers in oceanography Gustaf Ekman and Otto Pettersson, much of which was done in this area. Later there were surveys by Hans Pettersson, Nils Jerlov, and Börje Kullenberg in the 1930s and 1940s.

The animal communities on the soft bottoms of Gullmar Fjord were studied by Arvid Molander in the 1920s, while Torsten Gislén described fauna and flora on the hard bottoms. His works were based on direct observation by diving and underwater photography (1930).

In recent decades both the previous director of Kristineberg, Bertil Swedmark, and his successor, Jarl-Ove Strömberg, have collaborated with Alf Josefson and Tomas Lundälv on surveys designed to inventory the bottom fauna. Josefson's findings shed light on the structure and dynamics of the soft bottoms, while Lundälv has used underwater stereophotography to record events on marine hard bottoms. Lundälv's work has been successfully continued by Ib Svane, who has spent 12 years studying the relationship between reproduction patterns and population dynamics of various quantitatively important sea squirts. The surveys show qualitative and quantitative variations in the bottom fauna and underline the necessity for a long series of observations before the effect of various human activities can be assessed.

The fauna at Kristineberg offers interesting subjects for examination: studies of zoogeography, population dynamics, and ecology have particularly concentrated on various invertebrates. Among the many studies of polychaetes, may be mentioned the now classic work of Tycho Tullberg from the turn of the century and that of Gösta Jägersten and his pupils rather later. More recently, the biology of the bryozoans has been studied by Lars Silén. Lars Hernroth and Fredrik Gröndahl have done work on jelly fish, with particular reference to their complicated life cycle, while the population dynamics of mussels have been analysed by Björn Tunberg. The biology of different isopods is being dealt with by Jörnundur Svavarsson, and both crustaceans and isopods are included in projects currently being carried out by the director of the station, Jarl-Ove Strömberg.

Marine algal research at the station started at an early date, and special mention may be made of the studies carried out by Harald Kylin and his pupils both into cytology and physiology and into ecology. The taxonomy and systematics of green algae were dealt with by Carl Bliding and Johan Söderström, while Göran Michanek tackled the changes in the marine algal florae.

When a permanent position in marine plant biology was set up in 1980, it opened the way to a continuous program of research into marine plant physiology. Lennart Axelsson, who was the first holder of the post, concentrated his efforts on the physiology and ecology of sea weed, studying the differences among different groups of sea weed in a long-term project with Hans Ryberg of the Department of Physiological Botany and Inger Wallentinus of the Department of Marine Botany, both of Gothenburg University.

The studies involve measuring photosynthesis, both with oxygen production as a measure of the electron transport and pH changes as a measure of carbon fixation. Axelsson found that the three major groups of red, brown, and green algae show different patterns of adaptation to low carbon dioxide contents. Some brown algae use a buffer system to maintain electron transport for a limited time and produce oxygen when illuminated without access to an external source of carbon. Similar conditions may occur under natural conditions at low tide, when the algae are exposed to strong sunlight above the water level. When the algae are once more covered by sea water, the buffer system is charged, and the algae absorb both carbon dioxide and hydrogen ions from the water.

Marine creatures have in many cases been found particularly suitable for experimental analysis of general life functions. They present model systems that allow the scientist to tackle problems that may be difficult to study in the usual laboratory animals. In neurophysiology, for example, our understanding of the transmission of signals in nerve fibers and connections has largely been obtained using marine material. But for neurophysiological experiments, a knowledge of the structure of the nervous system is essential, and fundamental neuroanatomical projects have been carried out at the Kristineberg Station. Especial mention may be made of those completed early in this century by Gustaf Retzius, who investigated the fine structure of the nervous system, with results that even today provide an anatomic foundation for physiological analysis.

Transmission from one nerve cell to another is effected by neurotransmitters, which are of differing chemical structure in different parts of the nervous system. Some of the experiments of Ulf von Euler that led to his Nobel Prize in 1970 for the discovery of noradrenaline as a neurotransmitter were carried out at Kristineberg, where he and his colleagues studied the catecholamines adrenaline and noradrenaline in both fish and invertebrates. This transmission of signals has recently been studied by Rolf Andersson and Hans Elwing using a color change system in fish as a

Kristineberg Marine Biological Station, established in 1877, on the west coast of Sweden.

model. In neurophysiology, dogfish have been used by Sten Grillner and Peter Wallén to analyse the principles of the nervous control of rhythmically co-ordinated muscular movements.

Much of our knowledge of fertilization processes and of the processes that take place in the first stages of the development and differentiation of the embryo has have been obtained from studies of sea urchin eggs at the Kristineberg Station, mainly those carried out by John Runnström, Sven Hörstadius, and their many colleagues in the 1920s and 1930s, and those currently being conducted by Tryggve Gustafson. Gustafson is one of those scientists who return to the Kristineberg Station every year.

The different marine creatures available are also being used to examine many functional problems relating to, for example, body fluids, blood and blood cells, endocrine functions, and characteristics of different hard tissues in the skeleton and shell.

The bottom fauna in the deep waters is an extremely sensitive indicator of environmental changes. It is in this context that Odd Lindahl's studies of plankton community dynamics in relation to water exchange in Gullmar Fjord are of importance, as are Alf Josefson's and Rutger Rosenberg's studies of ongoing changes in the communities of bottom fauna in the Kattegat, the Skagerrak, and Oslo Fjord. It may be added here that a research group in ecological toxicology led by Märtha Swedmark and Åke Granmo, with support from the National Environment Protection Board, has recently examined the effect of various chemicals on marine organisms.

International interest in the Kristineberg Marine Biological Station is increasing, as is shown by the various international oceanographic conferences held there. Sweden must shoulder its responsibility for marine research in the Kattegat, the Skagerrak, and the North Sea. Sensible exploitation requires a knowledge based on marine research, which is also essential if constructive environmental policies are to be pursued. Continuation of activities at the Station assumes that pollution of the sea will be arrested, and the efforts that this requires must be based on marine research.

As it is acknowledged to occupy a unique place in marine research both in Sweden and internationally, it is a matter of honor that—in the words of Nobel prizewinner The(odor) Svedberg—Sweden's "most important scientific establishment for studies in this field" should enjoy sufficient financial support to be able to complete, under the wing of the Academy, its urgent research tasks.

The Mittag-Leffler Institute

One of the newer institutes of the Academy is the Mittag-Leffler Institute, originally a gift of Gösta Mittag-Leffler and his wife Signe. Mittag-Leffler was one of the leading figures in his field and the first Swedish mathematician of international standing. From 1881 he held the first chair in mathematics at the new Stockholm Högskola, later Stockholm University. Mittag-Leffler had outstanding pupils, including Ivar Otto Bendixson, Erik Ivar Fredholm, and Lars Edvard Phragmén. His work and the research tradition that was established during his time have had a vital influence on mathematics in Sweden.

In 1882 Mittag-Leffler started the journal *Acta Mathematica*, which quickly gained an international reputation and is still one of the world's leading mathematical journals. Mittag-Leffler's marriage to Signe Lindefors had made him a man of means, and he built a large villa in Djursholm, near Stockholm, and gathered a splendid mathematical library that included both classical and modern specialist literature. In 1919 he converted his real estate, his library, *Acta Mathematica,* and part of his personal fortune into a foundation for mathematics research, the Mittag-Leffler Foundation for Mathematics, which he then transferred to the Academy. His chief wish was "within the four Nordic countries and particularly in Sweden to maintain for the future and further advance the position in pure mathematics that these countries presently hold."

His own mathematical work involved the construction of single-valued analytic functions with prescribed singularities and analytic continuation of power series. His results are among the basic propositions in analytic function theory. He also made important contributions to the theory of linear differential equations with analytic coefficients.

Mittag-Leffler was the director of the Institute until his death in 1927, after which Torsten Carleman held the position until 1949. After an interregnum, activities resumed when the financial position improved towards the end of the 1960s, thanks in part to various donations to the Academy. The main building was renovated, living quarters were built for visiting research fellows, and the government set up a chair at Uppsala University, the holder of which was to be based with the Foundation. Lennart Carleson was appointed professor and director of the Foundation's Mittag-Leffler Institute. With this, a number of research projects began to take shape. The Academy's mathematics section and co-opted members representing Denmark, Finland, and Norway serve as the board of the Foundation.

The Foundation continued to publish *Acta Mathematica* and later also *Arkiv för Matematik,* the latter having previously been published by the Academy. The Institute has successfully taken charge of producing and distributing both journals, which are financially self-supporting.

The library, which as far as early literature in mathematics is concerned is unique, has grown into the finest of its kind in Sweden. The stock is constantly being renewed and at present totals 50,000–60,000 volumes. It is probably one of the largest private libraries in the world on the subject, and a great attraction to researchers.

Under Carleson's dynamic leadership, activities expanded rapidly to make the Institute an international research center of high scientific standard. The Mittag-Leffler Institute is also important in stimulating Swedish interest in new fields of mathematics. It has become established practice to draw up an annual research program according to the availability of qualified young Swedish researchers able to benefit from the new knowledge brought by the visiting research fellows. The Foundation has been able to place living accommodations at the disposal of most of them, which plays a great part in keeping teams together long enough to complete projects.

More than 50 scientists from different countries have visited the Institute every year. The fields of study have included harmonic analysis, probability, potential theory, complex analysis, and partial differential equations. The last-mentioned area was dealt with at length in 1975–1976 by Lars Gårding and Lars Hörmander. When Hörmander was director of the Institute in 1984–1986, the linear and nonlinear theory of this subject was the central theme. This had natural links with various applications. Hörmander's contribution was of great importance to the general work of the Foundation, and he was for many years the editor of *Acta Mathematica.*

Lennart Carleson was director of the Institute until 1984, when he was succeeded first by Lars Hörmander and then in 1986 by Dan Laksov. In recent years scientific activities have been organized in collaboration with experts from various universities. Research has been led with great success by Christer Kiselman and John Fornaess and has been concerned with algebraic geometry and several complex variable operator algebras.

The Abisko Scientific Research Station

The desirability of a scientific research station in the Torneträsk area was first argued in 1887 by the Lapland specialist Fredrik Svenonius, who considered that such a facility "would be of value to almost every branch of science, e.g., meteorology, zoology, botany, geology and physics." The matter was discussed in 1902 by the Society for Natural Sciences in Stockholm, which appointed a committee with Svenonius as its secretary and with a close connection to the Academy. It managed to acquire a provisional building on the shore of Vassijaure and received a few modest donations: that was the start of the Vassijaure Scientific Research Station.

It was hoped that work carried out there would give us a knowledge of the nature of the aurora, illuminate the particular climatic conditions at these high latitudes, shed light on the dynamic events behind the creation of mountain ranges, and contribute to our knowledge of the biological conditions for life in this subarctic environment. The Station attracted international notice, and when it burned down in 1910, the news was received with dismay. However, a private donation enabled a new building to be erected near Abisko on the southern shore of Lake Torneträsk.

The Abisko Scientific Research Station was completed in 1912, and the original building is still part of the Station's laboratory facilities. The Research Station was originally managed by the Vassijaure Committee of the Society for Natural Sciences, but in 1923 this Committee was reconstituted as an independent association, in which the Academy was represented. One of the members of the working committee of the association was the botanist Gustav Einar Du Rietz, who had started his phytogeographical research in the Torneträsk area. The geophysical studies were led by Bruno Rolf and included meteorological, hydrological, geomagnetic, and seismographic measurements. During the 1920s an average of 15 scientists did an annual total of 300–400 days' research at Abisko.

The Research Station was handed over in 1933 to the Academy, which kept activities going with the help of private donations as a necessary supplement to its own allocation and to a government contribution for meteorological measurements. When the Second World War broke out, activities were curtailed, and during the 1940s part of the Station was used by the military. Furthermore, it was now found that an area nearer to Kiruna would better meet requirements for its geophysical work.

The government commission set up at the instigation of Rolf Sievert to consider the future of geophysical research in Norrland proposed in 1944 that a geophysical laboratory should be built near Kiruna. The government remained cautious, but things began to move in 1948, when the Academy appointed its own interim committee for research in northern Norrland, on which Rolf Sievert was again the driving force. Provisional premises for geophysical research were built near Kiruna and the Abisko Research Station was refitted. Both establishments—Abisko and Kiruna—remained under the Academy's interim committee until 1952, when it was replaced by the Committee for the Academy's Research Stations in northern Norrland, under the chairmanship of Richard Sandler, the county governor. In 1956 the Riksdag made a grant to erect permanent buildings for the Kiruna Geophysical Observatory, to which were added funds from the Municipality of Kiruna and the Academy itself. The Observatory was opened in the summer of 1957. During the preceding ten-year period, the Academy's member Nils Amboldt had been in charge of measurement work at Kiruna, but Bengt Hultqvist was now appointed director, and in 1967 he went on to become professor of geocosmophysics at Umeå University. Activities expanded, and in 1973 the Kiruna Geophysical Institute became an independent research institute offering research training in collaboration with the University.

The refurbishment of the Abisko Research Station's instrumentation, which had been planned by the Uppsala botanist Gustaf Sandberg, made the Station more attractive, but larger and more modern premises were needed. In 1968 the Academy received government funds to deal with the accommodation problem, and by 1970 the new building was ready. When the Observatory at Kiruna became independent, the Abisko Research

Station acquired its own committee, with Ragnar Edenman, the former minister of education, as its chairman. Gustaf Sandberg served as director at Abisko from 1949 to 1973, when he was succeeded by the plant ecologist Mats Sonesson. Activities were expanded in the 1970s, and in 1982 a research post in meteorology and climatology was created, with Björn Holmgren as its first holder. A major extension of the research workers' living accommodation was planned in the 1970s and completed in 1985. With this, the Station was well provided with research premises and special laboratories, service facilities, and accommodation for permanent staff and researchers. Interest in the establishment is illustrated by the fact that in 1984 102 researchers were based at the Abisko Station, 183 participated in courses there, and 4,684 person-days were recorded.

The Abisko region is dominated by Torneträsk, an oligotrophic mountain lake, 70 kilometers long with a maximum depth of 180 meters. Within a relatively limited area, there are considerable variations in topography, geology, climate, flora, and fauna. The Caledonian bedrock offers a richly varied structure, and altitudes range from 342 meters above sea level on the surface of Torneträsk to almost 2,000 meters above sea level in the mountains. The climate shows a sharp gradient from the Atlantic type in the west to a more continental type in the east. The region also presents many different ecosystems, which reflect the diversity of the physical environment. The individual biotopic structures are usually simple, due to the climatic conditions at this northerly latitude (68°21′N, 18°49′E).

Regular photography of the aurora had already begun, under relatively primitive conditions, in the 1920s. At the same time, recording of the ionization of the atmosphere started, and in the 1930s the connection among the aurora, magnetic disturbances, and disruptions of radio communication was studied. After the Second World War, when geocosmophysical research had been transferred to Kiruna, Olof Rydbeck installed apparatus there for continuous recording of the different layers of the ionosphere, and Nils Amboldt took charge of the registration of magnetism. Continuous recording of the intensity of cosmic radiation also began during these years. These are activities that have expanded rapidly in recent decades at the Kiruna Geophysical Institute, where Bengt Hultqvist and his research group have been responsible for dramatic advances in ionospheric and magnetospheric research.

The Abisko Scientific Research Station in northern Lapland, with spectacular Lapporten (the Lapland Gap) in the background.

Glaciologists and geomorphologists were also quick to avail themselves of the opportunities offered at Kiruna. Several projects were designed to ascertain the nature of the Caledonian bedrock of the Norrbotten range and the history of the mountain chain, among them those of Alfred Elis Törnebohm, Oscar Kulling, and their colleagues. The tectonics of the Abisko area have been the subject of recent study by Maurits Lindström, while the dynamics of rock falls, avalanches, and other earth movements that contribute to slow changes in the mountain ranges have been examined by Anders Rapp and his colleagues. A number of investigators have used the wealth of material offered by the Torneträsk area to study deglaciation; these include Otto Sjögren, Carl Gustaf Holdar, and Olle Melander. Mention may also be made here of the study of the melting of Kårsa Glacier by Hans W:son Ahlman and his team and Carl Christian Wallén's work on the relationship between glacial melting and the region's climatic development.

During the 1950s and 1960s, Abisko was the base for extensive limnological work, particularly by Sven Ekman and Wilhelm Rodhe, who, together with their team, mapped out the physical, chemical, and biological conditions in a spectrum of lakes at different altitudes from shallow, humus-rich lakes in the birch belt at the level of Torneträsk to lakes in the alpine regions, 1,325 meters above sea level and with up to 40 meters transparency.

The climatic gradient from the local maritime type of climate at Riksgränsen to the continental type in the Östra Torneträsk area, like the altitude gradient, has given rise to a variety of types of vegetation, and much attention has been devoted to different plant associations, their geographical distribution and their ecological characteristics. During the years 1917–1921, Thore C.E. Fries was head of the Research Station, and his studies in floristics, taxonomy, and phytogeography paved the way for the ecological research taking place at Abisko today. His works on the alpine and sub-alpine vegetation of the Torne area also deal with the development of vegetation in Lapland after the glacial period. Einar Du Rietz, who led the program in biology in the 1920s, returned at different periods up to the 1950s to complete the study of the ecology of the mires and the classification of the alpine vegetation belts from the birch-belt region to the fall fields of the high alpine belt.

The marsh lands of the region have also been studied for comparison with similar types of landscape at northerly latitudes in the United States, Canada, and the Soviet Union. The nutrient-rich fens were examined in the 1960s by Åke Persson, while Mats Sonesson began to look at the structure, hydrology, mineral chemistry, and developmental history of the nutrient-poor bogs. There has since been extensive international research into the ecology of the subarctic mires; this formed an important part of the International Biological Program (IBP) initiated by the International Council of Scientific Unions (ICSU), which is represented in Sweden by the Academy.

The focus of the ecological research now in progress under Sonesson's guidance is the composition and dynamics of the different types of vegetation of the mountain regions. The subjects studied include the effect of climate on the growth and reproduction of different organisms at their distribution limits, where minor environmental changes have a dramatic effect on the organism's habitat, and also the adaptation of plants to various stress factors in the mountain region, such as air frost, ground frost, snow cover, scarcity of nutrients, etc. Investigations are also being conducted into the photosynthesis and respiration of different plant species and the dependence of these processes on different environmental factors.

The fauna studies at Abisko have concentrated on insects and, to a lesser extent, birds. The rich insect fauna of the sub-alpine birch belt has attracted many entomologists, and the numerous publications deal with taxonomy, ecology, and physiology. Over a 30-year period from the mid-1920s, Lars Brundin conducted his extensive studies of the taxonomy and ecology of beetles and non-biting midges, while Olle Tenov investigated the biology of autumn moths with particular reference to the dynamic relationship between climate, insects, and vegetation. One of the aspects studied is the relationship of the larvae to trees of different ages and to the local conditions and microbiology of the birch tree region.

In ornithology, mention may be made of Anders Enemar's studies of bird population density in the sub-alpine birch belt, a part of the faunistic inventory of Lapland organized by the Department of Zoology at Lund University.

The location of the Research Station, at a high latitude with the extreme light and climatic conditions that prevail at certain periods of the year, gives special opportunities for examining the effect of these external factors on the diurnal and annual rhythmic activity of various organisms, and biorhythmic studies of plants and animals have been carried out both in the field and under controlled laboratory conditions.

Since ionospheric and seismographic activities were transferred to the Kiruna Observatory, regular meteorological recording has continued at Abisko, as it provides information on climatic variations that is of relevance to phytoecological and geomorphological research. The establishment in 1982 of a research post in meteorology and climatology, with Björn Holmgren as its first occupant, has given a boost to climatic research in the Torneträsk area. There is at present a research project concerning the three-dimensional structure and dynamics of cold air masses in this arctic region and their drainage as cold airstreams into adjoining valleys; this is of interest to both meteorologists and biologists, among whom there has been Nordic collaboration on these questions.

The interest of scientists in the Torneträsk region, with its wealth of both abiotic and biotic variation, is reflected in the many scientific works that have resulted from the research carried out since the advent of the station at Abisko. Those published up to 1987 have been summarized in a bibliography by Nils-Åke Andersson, *et al;* it is based on Gustaf Sandberg's earlier bibliography (1964) and the supplement that Nils-Åke Andersson produced during the years 1967–1975. Altogether it gives information on some 1,700 publications, of which about four-fifths are in the biological sciences. The material illustrates both the increasing interest taken in the area by researchers, and how activities in the biosciences have gradually shifted from descriptive and taxonomic work to more functional studies.

The Research Station for Astrophysics

The astronomer and member of the Academy Yngve Öhman, who worked at the Academy's Stockholm Observatory at Saltsjöbaden, had developed solar research there using the narrowband filter that he had designed for the red spectral hydrogen line, the H-alpha line, of the solar atmosphere. In 1951 he and Per Olof Lindblad, mindful of the long hours of summer sunshine north of the Arctic Circle, explored the possibility of establishing a solar research station at the Academy's Scientific Research Station at Abisko. Beautiful pictures of solar prominences were obtained, but observations were hindered by the "clouds of sparks" resulting from the light-scattering effect of ice crystals and from swarms of midges at unexpected heights. Instead, the Academy turned southwards for its solar research.

The Swedish government had taken over the villa of the physician and author Axel Munthe at San Michele on the island of *Capri,* and when this was opened to scientists and artists, Öhman took his instruments there, where he found that atmospheric conditions were indeed favorable to solar research. At the solar eclipse of February 1952, he was in charge of a group of astronomers that the National Committee for Astronomy had sent to Capri, and when the observations produced encouraging results, the instruments were left there to form the nucleus of the research station that the Academy decided to set up. The chief instrument was a coronagraph with a grating spectrograph and an extra tube connected that was equipped with an H-alpha monochromator designed by Öhman and having a band width of only 1 Å.

As a result of the efforts of Öhman, the president of the foundation Centro Caprenze di Vita et di Studi, Edwin Cerio, became interested in setting up an observatory on the island. The Academy was allowed to rent a building up on Monte Solari, 480 meters above sea level, and also given free use of another building in Anacapri. The instrumentation was later augmented with a reflecting telescope and a 20-meter-long horizontal telescope with a grating spectrograph to study the magnetic fields of sunspots. The provisional premises were gradually replaced by more adequate buildings for instruments, laboratory work, and living accommodation. By the opening ceremony on June 18, 1961, the Academy buildings were on their own land, and there was room for further expansion. The last building was not completed until 1971; it houses a 60-centimeter reflector for stellar astronomy.

One of the problems that occupied Wargentin in his day was measuring the distance to the sun. The focus of studies now is physical events in the sun, our nearest star, which is middle-aged, and where hydrogen atoms fuse to create helium, and enormous amounts of energy are liberated. The solar atmosphere is so hot that hydrogen atoms lose their only electron and form a plasma consisting of electrically charged particles that is subject to the sun's strong and complex magnetic fields. This interaction is manifested in a varied "cloudy pattern" that can be seen and recorded with our modern solar telescopes and whose details can be spectroscopically analysed. Research at the Capri station concentrated on analyzing such phenomena as prominences and sunspots, occurrences that give us an idea of the physical events taking place in the sun's atmosphere.

Assisted by the fact that Öhman was chairman of the Commission on Solar Activity of the International Astronomical Union and by his wholehearted involvement in the International Geophysical Year 1957–1958, the Capri station developed during the 1950s into an important link in the global solar patrol network. Activities consisted primarily in analyzing phenomena that accompany the appearance of sunspots, i.e., flares, bright and dark surges, and eruptive prominences on the solar limb. Patrol activities formed the basis for more specific projects in that the continuous state of preparedness facilitated an exhaustive study of those phenomena that appear randomly and develop rapidly. Observations at the station benefitted from Öhman's many optical innovations. Velocity fields within active solar regions were subjected to comprehensive analysis, and together with Göran Hosinsky and Ulf Kusoffsky, Öhman conducted pioneering studies of the rotational effect of the solar gases. Much attention was also paid to the dark structures in the solar prominences, and Öhman's discovery of a group of prominences of extremely low temperature raised questions about the H_2 molecule in the solar atmosphere.

The continuous surveillance of the solar atmosphere also formed the basis for collaboration with geophysical research stations in Sweden and other countries in studying the connection between different forms of solar activity and the appearance of the aurora, which is produced by charged particles from the sun—electrons and protons—that are attracted by the magnetic field of the earth towards the areas around the poles. Under Öhman's direction activities on Capri expanded, and about 50 scientists worked there, most of them from university towns in Sweden.

In 1962 the European Space Research Organization (ESO) accepted Öhman's project of studying the sun in the ultraviolet region of the magnesium spectral lines. Kerstin Fredga had worked periodically at Capri during the years 1957–1967, and at the solar eclipse of 1961, she had taken part in Öhman's study of the electron temperature of prominences. When the Americans showed an interest in Öhman's project, Fredga participated on the Swedish side, and once collaboration with NASA and the Goddard Flight Center in Greenbelt began, she became responsible for the full scientific program. She supplemented the results she obtained at the three launches during 1965–1968 with studies at the station in Capri. When Öhman retired, solar surveillance was continued by Ulf Kusoffsky and Göran Hosinsky, who had been active as assistants since the mid-1960s.

In 1973 Arne Wyller was appointed professor of astrophysics at the Academy, and thus took over the leadership of the research station. He had been working for six years in astrophysics at the Barthol Research Foundation in Delaware (U.S.A.), and he continued to study the dynamics of sunspots with Hosinsky and Göran Scharmer from Stockholm University. Their studies, which were carried out both at Kit Peak National Observatory in the U.S.A. and with the aid of the satellite OSO-8 (1975), concerned the determination of the temperature of sunspot chromospheres.

For the future work of the station, Wyller developed a high-resolution spectrograph, the Ultravariable Resolution Single Interferometer Echelle Scanner (URSIES). The building of the spectrograph and the procurement of instrumentation for processing the data obtained were made possible by grants from the Wallenberg Foundation and the Natural Science Research Council.

While the solar telescope was gradually being improved, the astronomical climate in Europe was deteriorating. As pollution in the Mediterranean increased, it became apparent by the end of the 1960s that after its splendid pioneering efforts, the Academy's observatory would have to leave Capri. In 1968 representatives of seven European countries had formed the Joint Organization for Solar Observations (JOSO), with a brief to create a large, jointly owned solar observatory. Five more countries, Sweden among them, joined the project later. After test campaigns in about 40 different places, stellar astronomers in the United Kingdom and Denmark and solar physicists in West Germany and Sweden began to turn their attention to the Canary Islands, where there was already a Spanish observatory linked with the University of La Laguna in Tenerife. The German solar physicists chose the mountain peak of Teide on Tenerife, whereas the British and Danish stellar astronomers and the Swedish astrophysicists of the Academy of Sciences preferred the northerly peak of Roque de Los Muchachos on *La Palma,* the westernmost island of the Canaries. A total of 9,000 photoelectric recordings and 3,000 photographs of the sun had by then been taken, and a special test campaign was now organized to ascertain the best position for the Academy's observations on the edge of the extinct volcano, 2,350 meters above sea level. From 1974 onwards the planning and implementation of the transfer of the station from Capri to La Palma accounted for an increasing part of the staff's time. By 1977 the Swedish building project was in full swing.

On May 26, 1979, agreements were signed among the various parties from Denmark, Spain, Sweden, and the U.K. on the creation of the Spanish international observatory, *Observatorio del Roque de Los Muchachos.* For Sweden's part, the governmental agreement was signed by Ambassador Lennart Petri; the author had the honor of signing on behalf of the Academy of Sciences; and Arne Wyller signed the agreement for the Academy's research station. Separate agreements later gave Dutch and Irish astronomers access to the British observatories on La Palma. Finally, in 1983, the Germans signed an agreement to erect the sister observatory, *Observatorio del Teide* in Tenerife.

The shared resources that the two observatories represent are administered by *Instituto de Astrofísica de Canarias (IAC)* through its *Comité Científico International (CCI),* which is common to both observatories and has representatives from West Germany, the U.K., Denmark, Sweden, and Spain. The unique quality of this international observatory is the technical and financial autonomy of the participating partners, each of which is responsible for the construction, siting, and operation of its own telescope installations. Spain is responsible for the infrastructure, i.e., roads, electricity, telephone, hotels, and mess rooms; in return the Spanish astronomers have access to 20 percent of the observing time on the other countries' instruments.

The ceremonial opening of the observatories in the Canaries—on La Palma and Tenerife—took place on June, 28 and 29 1985. Heads of state of the seven participating countries took part in the festivities, together with administrative and research representatives, a total of 600 guests, with King Juan Carlos of Spain as the host. The Swedish observatory was opened by King Carl XVI Gustaf, accompanied by Queen Silvia; the instrumentation and the premises were demonstrated, and an exhibition was arranged to illustrate the development of solar physics, using material obtained from the solar station of the Academy of Sciences.

The new solar observatory consists of two parts: the 16-meter, five-storey solar tower and the two-storey laboratory building to which it is connected. The ground floor of the laboratory building contains rooms for spectrographs, data processing, photography, and an electronic workshop. On the upper floor, are a library, a conference room, research workers' accommodation, a canteen, and a kitchen.

The light-collecting optical system at the top of the solar tower was designed by Göran Scharmer. The solar image is formed by a double lens with an aperture of 48 centimeters, and the light is directed by two mirrors into the 16-meter-long vertical vacuum shaft. The URSIES spectrograph that was mentioned earlier with its magnetometer for the study of the magnetic fields in the solar atmosphere is installed in one of the spectrograph rooms on the ground floor. In the other is a Littrow spectrograph with an H-alpha filter.

When the telescope was tested for the first time on December, 1 1985, it proved to have a quality of image comparable with the best solar telescopes in the world.

Part of the cost of the solar observatory was met from the proceeds of the sale of the Academy's property in Capri. Grants from the Knut and Alice Wallenberg Foundation formed the bulk of the funds for erecting the solar tower and solar telescope, augmented by a donation from Arleen and Curtis Carlson, of Minneapolis. A donation by Mrs Margareta Dahlberg in memory of her father, Holger Crafoord, enabled the adjoining laboratory building, the Crafoord Laboratory for Solar Physics, to be built.

The collaboration within the IAC pays a scientific dividend on an investment that each country is able to contain within reasonable limits. It has radically expanded the research opportunities of Swedish astronomers, who have access in the organization to all the telescopes, for both solar and stellar astronomy. The British installations for stellar research include a 1.0-meter telescope, a 2.5-meter telescope (the Isaac Newton Telescope), and a 4.2-meter telescope (the William Herschel Telescope), together with extensive technical facilities and devices for remote operation. The Danish instrument is also intended for stellar research: a meridian circle for precise determination of the exact position of celestial objects (the Carlsberg Automated Transit Circle/CATC). The German equipment on Tenerife comprises a 0.6-meter telescope and a 0.45-meter telescope (the Gregory Telescope) for the study of solar magnetic fine structures; there is also a Nordic 2.5-meter telescope (Nordic Optic Telescope /NOT/), installed jointly by Denmark, Finland, Norway, and Sweden.

In 1986 Göran Scharmer installed a unique system that selects a series of sharp images of solar structures occurring in brief spells during moments of superlative seeing. Using this new system connected to the solar telescope, Scharmer obtained results of great significance for understanding the dynamics of the solar atmosphere. Part of the material was analysed in collaboration with Peter Brandt of the Freiburg Institute, West Germany, and Alan Title from Lockheed, Palo Alto, California (U.S.A.). Their discoveries are of importance to our understanding of convection in stellar atmospheres, of how magnetic fields are concentrated, and of how magnetohydrodynamic (MHD) waves are generated. Jan Högbom developed a new image reconstruction technique for the measurement of seeing distortion caused by inhomogeneities in the earth's atmosphere. In convection theory, finally, cooperation started with Åke Nordlund of the University of Copenhagen, who constructed a computer program describing the convection (granulation) in the solar atmosphere, together with the generation of all wave moods that can occur in a gravitational stratified gas with a magnetic field.

The Beijer Institute

During the 1950s and 1960s, it became steadily more obvious that the changes that man is causing in the global environment are threatening to have disastrous consequences if action is not taken to control them. The willingness to tackle the problems was expressed in the United Nations International Environmental Conference in Stockholm in 1972, for which René Dubos and Barbara Ward wrote the report on global environmental problems that has become known around the world under the title *Only One Earth.* The book was written at the request of the secretary-general of the conference, Maurice Strong, in consultation with many scientific correspondents.

During the year of the Stockholm Conference, the Academy marked its commitment to helping to fight the problems of our environment and resources by launching its international journal of environmental research, *Ambio.*

As the global problems of energy and the environment called for scientific assessment, the Academy naturally began early in the 1970s to devote much attention to these questions, both by arranging symposia and by taking part in international projects on the environmental problems of both developed and developing countries.

The Academy also improved its readiness in the energy and ecological sector by taking part in international scientific discussion of the global energy problem. At the Pugwash Conference of 1974 in Vienna, the wish was expressed for an international energy research institute designed for sophisticated scientific studies in the energy and ecology sectors. The importance was emphasized of linking this institute to a free and independent nongovernmental organization of international repute, to enable activities to proceed without undue influence from national and international pressure groups. For this reason the Academy of Sciences was considered a suitable parent organization.

The Academy was therefore well prepared on January, 28 1975 to receive the magnificent donation of Kjell and Märta Beijer for the establishment and operation of an international institute for energy and human ecology. The Beijer Foundation was set up to administer the funds provided by the donation, which enabled the Academy to build premises for the Beijer Institute alongside the main building in Frescati and to start activities for an initial period of eight years. As a guideline for its future activities, the Academy arranged, in November 1975, an international conference to discuss how the Academy's new institute could best fulfill its intended function within the network of global activities in progress in the fields of energy and ecology.

The planning group that the Academy later appointed put forward proposals concerning the general orientation and organization of the Institute. In accordance with these proposals, the Institute was established as one of the research institutes of the Academy, which guaranteed its independent status. In acknowledgment of its international character, an advisory international committee was created, and the director was selected from many foreign and Swedish candidates.

At the formal opening of The Beijer Institute, The International Institute for Energy and Human Ecology, on March, 14 1977, the newly appointed chairman, Professor Gordon Goodman, from Chelsea College, London, was able to outline the program for an introductory, exploratory period of activity, consisting of three partially coordinated research projects: energy risk management; environmental and international constraints on the European transition from oil to other energy sources; and improved energy utilization in developing countries. After serving as secretary of the *ad hoc* committee for the international planning symposium, Dr Lars Kristoferson became the first member of the permanent scientific staff of the Institute.

Within the first-named project category the Institute launched a comparative energy risk project that covered Sweden, France, Canada, the United Kingdom, the United States, and West Germany. The project included an evaluation of the utility of various energy risk studies and their effect on opinion-formers and decision-makers in society. This international comparison gave a sounder basis for energy risk assessments.

The second stage was eventually extended to cover the environmental and international consequences of the transition from oil to other energy sources, including consideration of economic and energy co-operation between oil-exporting and oil-importing countries, and also of current environmental questions concerning the use of fossil fuels. In the last-mentioned field, the Beijer Institute concentrated on environmental implications of and strategies for expanded coal utilization. An "east-west" working group, with representatives from several countries, was set up to look at questions of coal resources, combustion technology, and the environment. Mention may also be made of the study of the problems of the European gas market and the analysis of long-term global energy futures that were conducted in collaboration with the World Resources Institute. Activities on this front led to the establishment of an annex in England, the Beijer Institute Center for Resource Assessment and Management, at York University under the direction of Professor M. Chadwick. The program includes investigating the feasibility and cost of minimizing the effects of acidification in Europe.

The second working group, the "north-south" one, with representatives of developing countries intending to start or expand coal-mining operations (e.g., Bolivia, Botswana, Colombia, Tanzania, and Zimbabwe), concentrated its efforts on courses of action designed to alleviate the environmental impact of large-scale coal mining in developing countries. Within the scope of this project, the possibility is now being discussed of setting up a permanent working group to examine environmental and resource questions in developing countries (Mining Impacts Network, MINE).

In conjunction with the producer-gas experiment that it had concluded in the Seychelles, the Institute also started to study the feasibility of introducing producer-gas technology for providing developing countries with an alternative to oil. International collaboration began on projects in Sri Lanka and Nicaragua and a technical development program at the Royal Institute of Technology in Stockholm, the aim of which is to adapt producer-gas units to the different fuels available in different countries.

The most comprehensive projects fall within the third project area and concern energy use in developing countries. Together with Kenya's ministry of energy, the Beijer Institute drew up an energy study that became one of the largest and most exhaustive studies of this kind ever carried out in a developing country. The project was planned against the background of the fact that in Kenya wood and agricultural waste account for over 70 percent of total energy use, that fuel and wood dependence in rural areas are high, that oil dependence in the transport and energy sectors is almost total, that the fuel-wood situation is alarming, and that the exhaustion of timber supplies is so imminent that by the year 2000 it may be too late to take countermeasures in many parts of the country.

The Beijer Institute analysis resulted in a model plan of action that could if necessary be transferred to other developing countries with similar problems. The measures extended from energy plantations to improved stoves and industrial energy conservation.

The recommendations were agreed to by the Kenyan government, and the study was accepted as a scientific basis for deciding the country's energy policy. The study attracted international notice, and one of its consequences was that the Beijer Institute arranged a regional energy planning seminar for the South African Development Coordination Conference (SADCC), i.e., the economic organization for collaboration between nine states in southern Africa (Angola, Botswana, Lesotho, Malawi, Mozambique, Swaziland, Tanzania, Zambia, and Zimbabwe). The sequel to the work was that an analysis similar to that produced in Kenya was prepared in Zimbabwe in collaboration with that country's ministry of energy. The methodology of national energy planning was the most important area of research in the early 1980s, and the

Institute also devised a computerized energy planning system that integrates the fuel-wood supply of rural areas with the national energy picture. A similar utilization of this system was developed in both Zambia and Tanzania and was also used by the SADCC energy secretariat in Angola.

The Kenya project is illustrative of how the Beijer Institute plans its activities. During the work some 30 research scientists and 100 other persons were engaged in the project for varying periods, particularly at the Institute's annex in Nairobi. Field studies, technical analyses, and policy studies produced comprehensive material, and the results were published in a series of books and scientific articles.

In addition to finance provided for the project by the Kenyan ministry of energy and the Beijer Institute, assistance was given by a consortium of international aid bodies in Holland, West Germany, and the U.S.A. Some of the recommendations that were made led to long-term collaboration with the country's ministry of energy and the Dutch aid organization. A regional scientific center was established in Nairobi to deal with fuel-wood problems and agroforestry and to implement the fuel-wood and tree-planting projects proposed by the Institute. Out in the provinces, local centers were established in certain districts, such as Kakamega. These places were then the scene of the trial of new methods based on an analysis of technical, economic, social, and cultural factors. Following the evaluation of the project, which was carried out by the Dutch government, it has been expanded to cover more districts of Kenya.

The Beijer Institute has grown into a well-known center of research into energy, natural resources, and the environment. An internationally mobile network of scientists has facilitated the development of a flexible and cost-effective operation. This has enabled the Institute successfully to tackle a wider range of problems than many institutes with a larger permanent research staff.

To reflect the path that activities had taken during this first period of operation, the name was changed to The Beijer Institute, the International Institute for Energy, Resources and the Human Environment. To create a firmer foundation for future work on an international level, the Institute was reorganized in 1985–1986 as an independent foundation, created by the Academy and with an international committee appointed by the Academy. This was done after discussions among the Academy, the Beijer Foundation, and the Swedish government. The outcome of the discussions was that the Beijer Foundation undertook to support the Institute for a further five years with the assistance of a state grant. The core grant from Beijer constitutes 25 percent of the Institute's present total research allocation of approximately SEK 15 million per year.

Center for History of Science

Finally, it should be added that the Academy possesses a great many eighteenth, nineteenth, and twentieth-century instruments, accumulated over its 250-year history, as well as a large collection of letters, manuscripts, and other documents of considerable historical interest. The items that relate to the Swedish chemist Jacob Berzelius, including some of his personal possessions, form a separate collection, which is exhibited in the *Berzelius Museum*. This was first set up in 1898, moved to the Academy in 1914, and finally reorganized in 1973 on Academy premises near the main building. The museum illustrates the life and achievements of Berzelius, who served as permanent secretary from 1818 until his death in 1848 and had his residence at the Academy.

The main collection of instruments comprises about 5,000 items and forms the *Museum of the Exact Sciences*. Some of the instruments date back to the Academy physicist Johan Carl Wilcke (1732–1796), who was permanent secretary from 1784. Much of the collection originates from the Institute of Physics, established by the Academy in 1849 and active until 1922 (see above). Wilhelm Odelberg, the chief librarian of the Academy from 1959 until 1983, was instrumental both in the reorganization of the Berzelius Museum—together with the chemist Erik Jorpes—and in building up the Museum of the Exact Sciences. He served as director of both units until 1988.

Many of the scientific instruments are displayed in the *Center for History of Science* established in 1988 in connection with the coming 250th anniversary. This is located in the main building and was set up with the aid of generous donations. The Center is a research institute primarily for the study of the archives and other historical material of the Academy, including the instruments. The Center is also responsible for the Academy's Nobel Archives, of which material more than 50 years old is now available for research. The idea is to offer working space and service to guest scholars both from Sweden and abroad, and to arrange seminars and symposia in collaboration with the departments of History of Science at the universities in Stockholm and Uppsala.

Notes

This article on the research institutes is based on a series of monographs by Carl Gustaf Bernhard, all published and distributed by the Academy:

Abisko Scientific Research Station, KVA Bidrag XVIII (1989);
The Research Station for Astrophysics, ibid. XIX (1989);
The Kristineberg Marine Biological Station, ibid. XX (1989);
The International Institute for Energy Resources and the Human Environment: The Beijer Institute, in publication.
The Royal Swedish Academy of Sciences, ibid. XXI (1989). Complementing information can be found in this select bibliography:
Fries, R., "Några drag ur den Bergianska trädgårdens historia 1885–1914," *Acta Horti Bergiani,* Vol. 5 (1918).
Holmgren, N., "Kristinebergs Zoologiska station efter 75 år," *KVAÅ* 1953.
Hultqvist, B., "The Kiruna Geophysical Observatory, Sweden," *Nature,* Vol. 180 (1957).
Lindblad, Bertil, *Observatoriet i Saltsjöbaden* (Uppsala, 1931).
Idem., "Stockholms Observatoriums utveckling och verksamhet," *KVAÅ* 1954.
Pipping, G., *The Chamber of Physics,* KVA Bidrag XII (1977).
Odelberg, W., "Berzeliusmuseet i ny skepnad," *Svensk Naturvetenskap* 1973.
Ryberg, M., "Från Karlbergsvägen till Frescati: Bergianska trädgården under två sekel," *KVA Documenta,* No. 40 (1983).
Sonesson, M., "Abisko Scientific Research Station: Environment and Research," *Holarctic Ecology,* Vol. 2 (Copenhagen, 1979).
Swedmark, B., "Kristinebergs Marinbiologiska Station," *Svensk Naturvetenskap* 1974.
Westgren, A., "Vetenskapsakademiens forskningsinstitut för fysik," *Festskrift till Manne Siegbahn* (Uppsala, 1951).

Appendix

Permanent Secretaries

1739–1741	Anders Johan von Höpken
1741–1744	Jacob Faggot
1744–1749	Pehr Elvius
1749–1783	Pehr Wilhelm Wargentin
1784–1796	Johan Carl Wilcke
1796–1803	Daniel Melanderhielm
1803–1808	Carl Gustaf Sjöstén
1808–1811	Jöns Svanberg
1811–1818	Olof Swartz
1818–1848	Jacob Berzelius
1848–1866	Peter Fredrik Wahlberg
1866–1901	Georg Lindhagen
1901–1923	Christopher Aurivillius
1923–1933	Henrik Söderbaum
1933–1943	Henning Pleijel
1943–1959	Arne Westgren
1959–1972	Erik Rudberg
1973–1980	Carl Gustaf Bernhard
1981–1988	Tord Ganelius
1989–	Carl-Olof Jacobson

Publications

A. Closed journals and series:

Kongl. Swenska Wettenskaps Academiens Handlingar, Vols. I–XL, 1739–1779.

Kongl. Vetenskaps Academiens Nya Handlingar, Vols. I–XXXIII,
1780–1812.
Kongl. Vetenskaps Academiens Handlingar, 1813–1854.
Kongliga Svenska Vetenskaps-Akademiens Handlingar, Ny följd,
Vols. 1–63, 1855–1923.
Kungliga Svenska Vetenskaps-Akademiens Handlingar, Tredje
serien, Vols. 1–25, 1924–1948.
Kungliga Svenska Vetenskaps-Akademiens Handlingar, Fjärde
serien, Vols. 1–14, 1951–1974.

Bihang till Kungl. Svenska Vetenskapsakademiens Handlingar,
Vols. 1–28, 1872–1903.
Vols. 12–28 devided into:
Avd. 1. Matematik, astronomi, mekanik, fysik, meteorologi och
         beslägtade ämnen.
Avd. 2. Kemi, mineralogi, geognosi och beslägtade ämnen.
Avd. 3. Botanik, omfattande både lefvande och fossila former.
Avd. 4. Zoologi, omfattande både lefvande och fossila former.
Continued by *Arkiv för* . . .

Kungl. Svenska Vetenskapsakademien:
Praesidietal, 1739–1862
Åminnelsetal, 1743–1884.
Inträdestal, 1745–1792.
Ledamotsförteckning, 1766–1963 (continued by *Matrikel*).
Årsbok, Vols. 1–66, 1903–1968.

Årsberättelser om vetenskapernas framsteg, afgifne af KVA:s
embetsmän, 1821–1851.
Öfversigt af Kongl. Svenska Vetenskapsakademiens förhandlingar,
Vols. 1–59, 1844–1902.
Acta Horti Bergiani, Vols. 1–20, 1890–1967.
Arkiv för astronomi, Vols. 1–5, 1950–1974.
Arkiv för botanik och phytopaleontologi, Vols. 1–14, 1886–1901.
Arkiv för botanik, Vols. 1–33, 1903–1948. Serie 2, Vols. 1–7, 1963–74.
Arkiv för fysik, Vols. 1–40, 1949–1974.
Arkiv för geofysik, Vols. 1–5, 1950–1974.
Arkiv för kemi, Vols. 1–32, 1949–1974.
Arkiv för kemi, mineralogi, geognosi m.m., Vols. 1–9, 1886–1901.
Arkiv för kemi, mineralogi, geologi, Vols. 1–26, 1903–1949.
Arkiv för matematik, astronomi, fysik, meteorologi m.m., Vols.
1–16, 1886–1901.
Arkiv för matematik, astronomi, fysik, Vols. 1–36, 1903–1949.
Arkiv för mineralogi och geologi, Vols. 1–5, 1949–1974.
Arkiv för zoologi och zoopaleontologi, Vols. 1–11, 1886–1901.

Arkiv för zoologi, Vols. 1–42, 1903–1950. Serie 2, Vols. 1–23, 1950–1974.
Avhandlingar i naturskyddsärenden, Vols. 1–22, 1938–1966.
Ekonomiska annaler, Vols. 1–8, 1807–1808.
Samling af rön rörande landtbruket, Vols. 1–5, 1775–1788.
Skrifter i naturskyddsärenden, Vols. 1–53, 1919–1969.

B. *Current journals:*

Acta Mathematica, since 1882–
Acta Zoologica, since Vol. 62, 1981–
Ambio, 1972–
Arkiv för Matematik, 1949–
Chemica scripta, 1971–
Physica scripta, 1970–
Zoologica scripta, 1971–

C. *Current series:*

Bidrag till Kungl. Svenska Vetenskapsakademiens historia, Vol. 1, 1963–
Kungl. Svenska Vetenskapsakademien: Levnadsteckningar, Vol. 1, 1869–
Kungl. Svenska Vetenskapsakademien: Matrikel, 1964–
Kungl. Svenska Vetenskapsakademien: Porträttmatrikel, 1971–
Documenta, 1972–

Index of Names

The Academy's first *ex libris.*

15 Herr **Edvald Ribe**, Kongl. Archiater och Præses i Kongl. Colleg. Med. 173
                                           d. 8
                       vald d. 9 Junii, år 1739.
                       Har ingifvit 4 Rön til Academiens Handl.
                       var Præses för Månad Jul. Aug. Sept. 1740.
                       Herr Mörck har hållit åminnelse Tal öfver Honom.

16 Herr **Johan Julius Sahlberg**, Amiralitets Apothekare i Stockholm. 175.
                                             d. 18
                       vald til Ledamot d. 16 Junii, år 1739.
                       Af Des til Acad. ingifne Rön, äro 11 införde i Handl.
                       Var Præses för Månaderna Apr. Mai. Jun. 1745.
                       åminnelse Tal öfver Honom är hållit af Herr C.F. Ribe.

17. Herr **Jacob Faggot**, öfver. Directeur vid k. Landtmäteri- Contoiret. 177
                                             d. 28
                   var några år k. Vetensk. Academiens Secreterare, har varit 2 gånger
                   Præses, ingifvit 20 Rön, och jämväl i öfrigt gjort sig känd af det Allmänna
                   och af Academien mycket förtjent.
                   åminnelse Tal öfver honom är hållet af Herr Nicander.
                   Secreterare i Vet. Acad. fr. 1742 års början til 1744 års slut.

18. Herr **Lorentz Christopher Stobée**, General Major af Fortification, Landshöfding 172
                                             d. 3
                       öfver-Commendant och Riddare. Blef Ledam. år 1739,
                       men uteslöts, år 1747.

19. Herr **Gilbert Scheldon** Öfverste Lieutnant vid Flottornas Con- 1794
                                             d. 20
                   structions Stat
                       åminnelse tal öfver honom är hållit af Zjöstén

20. Herr **Lars Roberg**, Medicinæ Professor i Upsala. 174
                       Vald til Ledamot, år 1739, men
                       finnes ei hafva ingifvit något Rön.

21. Herr **Samuel Klingenstierna** Stats Secreterare, Ridd. af Nordst. Orden. + 176
                                             d. 26
                   vald til Ledamot vid k. Academiens Stiftares första Sammanträde,
                   år 1739. Var Præses i Jul. Aug. Sept. 1755. Uti Academiens Handlin-
                   gar finnas gaf Des arbeten. Öf honom värdigt åminnelse Tal är hållet
                   af Herr Strömer.

22. Herr **Carl Fredric Nordenberg**, adlad Nordenschöld. öfverste + 177
                                             d. 19
                       vid Fortificationen, Ridd. af S.O.;
                   var Præses i Apr. Maj. Jun. 1758. Har ingifvit 6 Rön.
                   åminnelse Tal öfver honom är hållet af Herr General Majoren
                             von Arbin.